Marine Climate of Russian Coastal Territories

Andrei Vyacheslavovich Tarasov
Rofail Salykhovich Rakhmanov

Marine Climate of Russian Coastal Territories

Public Health Aspects of Biological Adaption

Springer

Andrei Vyacheslavovich Tarasov
Department of Pediatrics and Preventive Medicine
Immanuel Kant Baltic Federal University
Kaliningrad, Russia

Rofail Salykhovich Rakhmanov
Professor of Hygiene Department
Privolzhsky Research Medical University
Nizhny Novgorod, Russia

ISBN 978-3-031-30950-2 ISBN 978-3-031-30951-9 (eBook)
https://doi.org/10.1007/978-3-031-30951-9

This Springer imprint is published by the registered company Springer Nature Switzerland AG
The registered company address is: Gewerbestrasse 11, 6330 Cham, Switzerland

Acknowledgments

The authors would like to thank their families for their support. We would also like to extend our thanks to the staff of I. Kant Baltic Federal University: Maxim Demin and Anna Silnichnaya for organizing this project; Elena Boyarskaya, Anna Bryushinkina, and Lyudmila Boyko for helping to translate this book into English.

This monograph presents a comprehensive analysis of the process of adaptation and adaptation capacity of population to the marine climates of Russia's North, the Black, Caspian, and Baltic Sea coastal regions. The authors explore different biological reactions to adaptation and offer their classification depending on the type of climate. Special attention is given to the prevention of prenosological conditions and diseases typical of the coastal areas analyzed. Evidence suggests that the human body has a remarkable capacity to adapt to a range of climatic and weather conditions through a variety of adaptation reactions.

The monograph was written based on the material of the post-doctoral thesis prepared under the supervision of Professor Rofail S. Rakhmanov.

Contents

About the Authors

Andrei Vyacheslavovich Tarasov, MD, PhD, Associate Professor at the Department of Pediatrics and Preventive Medicine, Immanuel Kant Baltic Federal University, Kaliningrad, Russia. Russian scientist, epidemiologist, hygienist. Scientific interests: study and correction of human adaptation and acclimatization mechanisms in the Baltic region.

Rofail Salykhovich Rakhmanov Honored Doctor of Russia, MD, Professor of Hygiene Department, Privolzhsky Research Medical University, Nizhny Novgorod, Russia. Russian scientist, epidemiologist, hygienist. Founder of the academic school of the reduction of tropical diseases during adaptation and acclimatization; founder of the research field of targeted foods products. Scientific interests: study and correction of mechanisms of human adaptation and acclimatization.

Chapter 1
Introduction

The problem of adaptation to climate and the prevention of diseases related to it have always been high on the agenda of the Russian government, and particularly so during the development of Russia's remote territories – the Arctic and the Far East. Recent research has shown that residents of these regions are prone to climatic stress caused by the adverse effect of the climate on the body regulatory system. Evidence suggests that these prenosological conditions increase morbidity among those who have relocated to these regions of Russia from other parts of the country.

Physiologists, hygienists, general practitioners, and specialists from other fields of medical science have proven valuable information on the adaptation capacity of the body to extreme environmental conditions. The works of academician Kaznacheev [1], Koscheev [2], Novozhilov [3], Azhaev [4], and Chvyrev [5] explore the critical role of climatic adaptation in the normal functioning of the body systems.

Being the most important component of the natural environment, climate affects economic activity, everyday life, living conditions and sanitation, and last but not least, health. The occupational health and physiology of people living in extreme climatic conditions have always been an important area of medical research. In the Soviet Union, since the 1930s–1940s, this topic was of particular importance. This was the time of the industrialization of the country and the development of arid and previously unpopulated areas of Central Asia. In the 1950s–1970s, research into the impact of climate on human health took on an added importance due to the development of the oil and gas industry and the excavation of mineral resources in the Arctic. The works of Adolf (1952), Galanin (1969) and others [6, 7], Malyshev (1963), and Sapov and Novikov (1984) have made an important contribution to the study of adaptation capacity to extreme climatic conditions of Russia's south and north [8, 9]. To date, these works have not lost their relevance.

It has been established that after moving from one region of the country to another, people have to adapt to the new climatic conditions not only physiologically but also psychologically and socially [10]. Researchers have established correlations between environmental factors and the health of different population

A. V. Tarasov, R. S. Rakhmanov, *Marine Climate of Russian Coastal Territories*,
https://doi.org/10.1007/978-3-031-30951-9_1

groups. Special attention has been given to the development of the principles and methods of prenosological diagnostics and the identification of the adaption processes of the body [11].

The study of the inextricable connection between human health and the environment, weather, and climatic factors began in Antiquity. Hippocrates and Avicenna wrote about an obvious connection between health and climate. In The Canon of Medicine, Abu Ali ibn Sina (Avicenna) showed that "the change of the season produces some kinds of diseases in any climate" (1020).

In Russia, the development of climatology as a science dates back to the eighteenth century. In 1753, the famous Russian scholar Lomonosov published The Word about Air Phenomena, in which he pointed out the importance of weather for human health.

The founder of Russian climatology was Alexander Voeikov. In his seminal works, he showed the interdependence of various weather phenomena, explored the impact of solar heat, barometric pressure, air circulation, and geology on human health. In 1893, he published a study on climate and its role in causing and preventing diseases. Luminaries of Russian medicine – Mudrov, Ostroumov, Botkin, and Zakharyin – often used the healing capacity of nature for improving human health. As early as the 1880s, Ostroumov argued that simple changes in lifestyle, for instance, moving to the countryside, swimming, and walking could have a beneficial effect and improve the state of the nervous system. Botkin was the first to study the healing properties of the Crimean climate and recommended the South Coast of Crimea as a climatic resort. In his lectures, Zakharyin [89] argued that there were no diseases in the incidence of which climatic and meteorological conditions would not play an important role. In the middle of the twentieth century, a detailed summary of medical and meteorological observations was made by Assman [90].

A new area of research – bioclimatic pathology – began to form in the twentieth century. It studies the reactions of the human body to weather, climate, geophysical, and spatial factors. Russian scientists [91–94] continued to explore the physiological and pathological effects of weather. They identified a new group of diseases which they called "meteotrophic," that is, diseases associated with the exposure to climatic, meteorological, and geophysical factors, and offered their classification: (1) diseases caused by thermal stress; (2) diseases caused by ultraviolet radiation; (3) infectious diseases; and (4) annually recurring diseases (seasonal diseases).

The past few decades were the time of the formation of clinical meteopathology as a branch of science. Prominent Russian scientists, Danishevsky, Fedorov, Voronin, and Grigoriev, have made a significant contribution to its development.

Climatology has also been developing very rapidly. Climatologists cooperate with other researchers working in physics, biology, and medicine [12].

Climatic environmental factors can affect humans and have pathological and health-promoting, stable and unstable, and direct and indirect consequences [13–16]. Long-term exposure to adverse climatic factors defines morbidity patterns; once people have adapted to unfavorable factors, prenosological conditions develop [17, 18]. Climatic features of a particular region (territory) determine thermal comfort or stress and daily energy needs. Thus, a healthy individual's exposure to low temperatures increases the need for energy and causes the so-called "cold thermogenesis" to ensure a thermal balance with the environment. Hot climate (depending on air humidity – the amount of water vapor in the air):

- In dry climate conditions, leads to significant energy losses of heat due to increased sweating (evaporation of 1 liter of sweat consumes 2436 kJ of energy or about 58 kcal/1 gram of sweat); that is, it affects nutrient metabolism, thus disrupting water and electrolyte body balance.
- In humid climate conditions, disrupts sweating, leading to body overheating at lower than dry climate temperatures.

Medical climatology is a science that studies the characteristics of climate and weather in terms of their impact on the body. Medical climatology includes several branches:

- Medical geography, which studies spatial patterns of the spread of diseases in different geographic zones.
- Climatophysiology, which studies the effects of a combination of different climatic and weather factors on the body, as well as changes that occur after moving from one climatic region to another.
- Climatopathology, which studies the dependence of various pathological conditions on climatic and weather factors.
- Climatotherapy, which is the use of the beneficial properties of a more favorable climate for the prevention and treatment of patients with various diseases.

In recent years, two other areas in medical climatology have been actively developing:

- Assessment of bioclimatic comfort/discomfort of specific territories (dealing with medical and social aspects of territorial development and population habitation) and of people's working capacity (defining work and rest conditions in the open air, providing personal protective equipment).
- Assessment of climate change-borne health risks causing increased morbidity and mortality of the population exposed to heat or cold.

Chapter 2
Climate and Weather: Impact on the Body

2.1 Definition and Classification: Biological Reactions of Adaptation to Different Types of Climate

Being closely interconnected, environmental factors influence the body in a complex manner. Depending on the type, their effect on the body can be different. Two key notions, *weather* and *climate*, are traditionally used for a comprehensive assessment of environmental factors.

Climate is a long-term weather regime, one of the main geographic characteristics of a particular territory [19]. Climate-forming factors include latitude and longitude (determining the magnitude of solar radiation and radiation balance), atmospheric air circulation, the type of terrain, and the nature of the underlying surface [20].

Climatic conditions are formed by the movement of air masses, the direction and the speed of the wind, which contributes to the equalization of barometric pressure on the ground since air moves from the areas of high barometric pressure to those with lower pressure. In summer, due to the more intense heating of land, the barometric pressure above it is lower than that above water. In winter, due to the cooling of the land surface, the barometric pressure is higher above land than that above water. This contributes to the intensity of the movement of air masses. In summer, humid air moves from sea to land, and in winter, dry air moves from land to sea. These winds are called monsoons [21].

Coastal winds – breezes – travel from sea to land during the day and from land to sea at night. Another type of wind, the so-called valley wind, blows from the sea toward and along valleys during the day. Mountain winds move air masses in the opposite direction.

Being the most important component of the natural environment, climate affects economic activity, everyday life, living conditions, health, and last but not least, the structure and the rate of morbidity. The spreading of pathogens and their carriers

A. V. Tarasov, R. S. Rakhmanov, *Marine Climate of Russian Coastal Territories*,
https://doi.org/10.1007/978-3-031-30951-9_2

depends on the type of climate. It is now known that the geographical distribution of many diseases is associated with climate. Therefore, climatic conditions should be taken into account during the elaboration of hygiene recommendations for civil and industrial construction, the production of clothes and footwear, optimization of work and rest schedules, the improvement of diet and nutrition, and the prevention of the onset and exacerbation of various diseases [22].

Climate is essential for a wide variety of processes. There are several applied classifications of climate. According to the classification used in construction, the territory of the Commonwealth of Independent States (CIS) is divided into four climatic zones based on the average temperatures of January and July: zone I – cold climate; zone II – moderate climate; zone III – warm climate; and zone IV – hot climate. This classification is widely used in urban and settlement distribution planning, the spatial orientation of buildings, and also for determining the thickness of walls, the type of heating, the size of windows, the depth of water pipes, etc. However, this classification does not take into account the impact of climate on the human body.

There is a medical classification of climate, which is used in the system of public health care and in the sanatorium-resort business. According to this classification, all known types of climate in Russia can be divided into two groups – marine and continental. It is noteworthy that in medical practice all types of climate are often divided into mild and harsh. A warm climate with low-temperature amplitudes and minor annual, monthly, and daily fluctuations of meteorological parameters is considered to be mild. Forest-covered territories of Central Russia and the climate of the southern coast of Crimea are examples of mild climate. This type of climate requires only a short period of physiological adaptation.

A harsh climate is characterized by considerable daily and seasonal fluctuations of meteorological parameters that negatively affect the adaptive mechanisms of the body [23]. The Arctic zone of Russia, mountainous regions, and hot deserts and steppes are typical examples of this type of climate. The climate of the northern territories of Russia, located in the permafrost zone, is particularly unfavorable: it is characterized by low air temperatures, high relative humidity, strong winds, and a lack of solar radiation.

These climatic conditions cause disorders in thermoregulation and hemodynamics, lead to an increase in basal metabolism, hypersecretion of the stomach, nervous system disorders (increased inhibition), the weakening of conditioned reflexes, decreased work performance, and sleep disorders (particularly during the polar day). Low temperatures in combination with high humidity lead to increased incidence of respiratory diseases, rheumatism, and diseases of the peripheral nervous system such as radiculitis, neuritis, and myalgia among others.

The hot climate of deserts and steppes is characterized by hot summers, a sharp variation of daily temperatures, low humidity, and excessive solar radiation. These conditions can cause overheating, heat and sunstrokes, electrolyte and water imbalance, decreased basal metabolism, hemodynamic disorders (capillary dilation, low blood pressure, and tachycardia), disorders of the gastrointestinal tract (loss of appetite, excessive thirst), increased incidence of intestinal infections (dysentery, typhoid, paratyphoid, cholera, etc.), and food bacterial poisoning due to rapid spoilage of food, as well as the rapid growth of bacteria and fast reproduction of insects-carriers of infections. There is a greater risk of skin cancer due to excessive ultraviolet radiation (especially in people having pale, low-pigmented skin), urolithiasis due to mineral metabolism disorders caused by consuming large quantities of highly mineralized drinking water, and cataracts due to excessive exposure to infrared light. In addition, there is a decrease in work productivity.

The high-altitude climatic zone is located 2000 m above sea level or higher. It is characterized by lower barometric pressure, clean air, intense solar radiation, and low oxygen partial pressure. These conditions can cause hypoxia, which stimulates compensatory erythropoiesis. The depth and rate of respiration change: it becomes less frequent and much deeper than the norm. The oxyhemoglobin dissociation curve also changes, the process of oxygen adhesion and separation is accelerated, and thermoregulatory processes are compromised. People living in this climatic zone often experience hyper-dryness of mucous membranes of the eyes and upper respiratory tract and abnormal sensitivity to or intolerance of light.

Territories that have marine, mountain, forest, and steppe types of climate are traditional recreation zones as they boast favorable meteorological conditions for improving health.

Biological reactions of adaptation or adaptation to weather and climatic conditions have become an increasingly important area of research since they are vital for the existence of plants, animals, and humans in different places of the globe [12].

2.2 Weather, Types of Weather, and Their Impact on Public Health

Weather is the physical state of the atmosphere at a particular point in time. Weather is usually characterized by the intensity of solar radiation, atmospheric electricity, temperature, humidity, air pressure, wind speed and direction, and atmospheric precipitation [19]. Unlike climate, weather is an unstable set of meteorological conditions that may change several times during 1 day.

Weather can be defined as a holistic natural system, characterized by several interrelated and interdependent meteorological parameters and phenomena.

A cyclone, a low-pressure air mass of about 2500–3000 km in diameter, forms in warm air and moves rotating around a low-pressure center. Cyclones bring unstable weather with significant fluctuations in barometric pressure, and temperature as well as high humidity, precipitation, and air conductivity. An anticyclone is a weather system having high barometric pressure. Anticyclones usually have a diameter of about 5000–7000 km and bring stable but not necessarily fine and sunny weather.

Recently the problem of the impact of weather on human health has become particularly acute due to the continuing climate change, and an increase in the number of meteorological anomalies – deviations of meteorological characteristics from their average values in time or space. Much of the literature emphasizes the need to study the negative health response triggered by weather factors. Previous research into these medical conditions focused on disorders of physiological adaptation to climate, consequences of acute meteorological stress, and the adverse effects of weather. The negative impact of climate and weather on health is not so much related to the values of meteorological parameters. Rather it is caused by their abrupt changes, which increase pressure on the endocrine and central nervous systems regulating homeostasis and lead to the desynchronization of internal biological rhythms.

It is now well established that weather can influence health and the physiological functions of the human body directly and indirectly. The direct impact is manifested in body heat transfer, may cause heat stroke and hypothermia, and may result in the weakening of the immune system, a higher incidence of respiratory diseases, diseases of the peripheral nervous system, neuritis, radiculitis, neuralgia, etc. The indirect impact of weather is mainly associated with the aperiodic changes in weather conditions, which do not correspond to the normal biological adaptive rhythms of the body: diurnal, monthly, annual, and heliobiological, which are conditioned by the 11-year solar magnetic activity cycle.

Several classifications of weather are widely used in medical climatopathology. Rumyantsev and his co-authors have proposed a classification based on the estimation of the mean weighted skin temperature. The researchers distinguish between (a) comfortable weather, (b) cold weather, and (c) warm weather.

There is another classification, in which weather is graded in several types depending on the degree of exposure to it and the effect produced. The most favorable weather types are defined as "comfortable," "warm," and "cool." Two weather types are "very warm" and "cold." Unfavorable weather can be of four types – "hot," "very cold," "very hot," and "extremely cold."

In the Fedorov–Chubukov system, the weather is divided into three main types – frost-free weather, zero-temperature weather, and sub-zero weather. In total, Fedorov and Chubukov singled out 16 types of weather as follows:

Frost-free weather:

1. Sunny, very hot, and very dry.
2. Sunny, hot, and dry.
3. Sunny, moderately humid, and wet.
4. Cloudy during the day and partly cloudy at night.
5. Sunny, moderately humid weather with cloudiness during the day.
6. Cloudy weather without precipitation.
7. Rainy (overcast with precipitation).
8. Very hot and very humid.

The zero air temperature weather:

9. Cloudy during the day (0 °C transition days in cloudy weather).
10. Solar (with the sun out).

Sub-zero weather:

11. Slightly frosty.
12. Moderately frosty.
13. Significantly frosty.
14. Very frosty.
15. Brutally frosty.
16. Extremely frosty.

According to Azhitsky, weather types are classified as follows:

- The "favorable weather" group is characterized by minor fluctuations of 14 main meteorological components, with daily variability of atmospheric pressure not exceeding 4 mbar (1000 mbar = 750.1 mm Hg), air temperature of 2 °C, windsock wind speed under 3 m/s, and relative humidity in the range from 55% to 85%. As a rule, favorable weather is well tolerated by patients and makes it possible to deliver all types of climatic therapies. This group includes weather classes 2, 3, 4, 5, 9, 10, and 11, according to the Fedorov–Chubukov classification. In class 2 weather, in order to avoid overheating, morning and afternoon hours are recommended for sunbathing, physical therapy, and sports activities.
- The "relatively favorable" weather group is characterized by the changeability of meteorological factors, with daily atmospheric pressure fluctuations of 5–8 mbar, air temperature 3–4 °C, relative humidity dropping below 55% or rising above 85%, and windsock wind speed exceeding 4 m/s. This weather group includes weather classes 1, 6, 8, and 12. The relatively favorable weather is well tolerated by healthy people and satisfactorily tolerated by many patients. However, in some cases, meteopathic reactions may occur, which may require drug therapy and a certain regimen. Strong winds and precipitations impede climatic therapies.
- The "unfavorable" weather group is characterized by sharp changes in the smooth course of meteorological factors, with the daily variability of atmospheric pressure over 8 mbar, and air temperatures above 4 °C. The group of unfavorable weather includes weather classes VII, XVI, XIII, XIV, and XV, according to Fedorov–Chubukov. It also includes the weather conditions of the "relatively favorable" group, but those are accompanied by a strong wind (over 9 m/s), accompanied by thunderstorms, fogs, hails, snowstorms, dust storms, and other menacing elements. In adverse weather, many patients experience meteopathic reactions. In such cases, special therapy and regimen are needed. Active forms of climatotherapy are not prescribed in adverse weather conditions.

The weather classification by Rusanov distinguishes: (1) clinically favorable weather characterized by an interdaily increase in barometric pressure and air temperature; (2) clinically less favorable – with an interdaily decrease in pressure and temperature; (3) clinically unfavorable – with an interdaily decrease in pressure and increase in temperature; (4) extremely unfavorable – with an interdaily increase in pressure and decrease in temperature. It was found that the incidence of pathological reactions in patients with cardiovascular diseases doubles, triples, and quadruples during the weather of the second, third, and fourth types, respectively, compared with type one.

To assess the weather for meteorological prophylaxis, a medical weather classification is proposed, called a morpho-dynamic classification. In this classification, the whole variety of weather conditions is broken up into four medical types, in meteorological terms categorizing Weather Types I and II as favorable, and Types III and IV as unfavorable. Weather Types I and II are formed mainly within anticyclonic atmospheric circulations. Usually, they are marked by stable, slightly cloudy weather without sharp disturbances in the normal diurnal course of meteorological variables and any pronounced variability of bio-geophysical values. Weather Types III and IV are formed mainly during cyclonic atmospheric circulations. Type III weather is marked by a disturbance of the daily course and a significant variability of the main meteorological parameters. Type IV weather is characterized by pronounced atmospheric fronts, disruption of the diurnal cycle, and sharp fluctuations in meteorological and geophysical factors.

Features of Type I weather. The Earth's surface and the lower troposphere are the areas of increased pressure. There are no atmospheric fronts, ascending vertical currents are weak, and wind shear is of medium or weak strength. No significant fluctuations in air temperature and relative humidity are observed. Atmospheric pressure fluctuations do not exceed 1 mbar per 3 h; wind speed stands at 0–3 m/s. Atmospheric oxygen content changes insignificantly within ±5–10 g per 1 kg of air in the 6–12 h period. The intensity of the atmospheric electric field near the Earth's surface is close to normal. No dangerous natural phenomena are observed. This weather type holds for 31–42% of the year.

Features of Type II weather. At the Earth's surface and in the troposphere, the atmospheric pressure changes slightly, with vertical air currents being weak. Frontal passages may occur, while air properties do not change significantly. The temperature and relative humidity of the air are within the seasonal and daily norms, with the wind speed in the range of 4–10 m/s. Fluctuations in the oxygen content are within ±10–15 g per 1 kg of air. The intensity of the atmospheric electric field is close to normal values. Thunderstorms and short-term rains – or snow in the winter period – are to be expected. This weather type is typical for 29–52% of the year.

Type III weather is characterized by the formation of cyclones with pronounced frontal passages and ascending vertical air currents. Air temperatures can shift by 10–20 °C in 6–12 h, relative humidity by 20–40%, and atmospheric pressure by 3–4 mbar within 3 h. Wind speed can increase up to 10–16 m/s. Oxygen content fluctuations stand at ±15–20 g per 1 kg of air. The voltage of the atmospheric electric field differs markedly from normal values; geomagnetic disturbances are possible.

Type IV weather is characterized by the active formation of cyclones with pronounced atmospheric fronts and ascending air currents. Natural hazards and large-scale natural hazards may occur – thunderstorms, squalls, hurricanes, downpours, snow and dust storms, etc. On average, Type IV weather holds for 5–8% of the year. The morpho-dynamic classification is used to develop medical weather forecasts.

Table 2.1 The degree of variability of meteorological parameters

Degree of variability	Degree of expression	Possible negative effects on health			
		Barometric pressure, mbar	Air temperature, °C	Relative humidity, %	Oxygen concentration, g/m³
I	Very weak	±0.25	±2.5	±10	±2.5
II	Moderate	Twice as high as degree I			
III		4 times more than degree I			
IV	Strong	8 times more than degree I			
V		8 times or more than degree I			

In recent years, the classification of weather developed by the Central Research Institute of Balneology and Physiotherapy has become more widespread. It includes seven basic types of weather: (1) stable neutral weather; (2) unstable weather changing from the neutral to spastic type; (3) spastic weather; (4) unstable spastic weather with a hypoxic component; (5) hypoxic type; (6) unstable hypoxic type with spastic weather components; and (7) spastic type changing into the neutral one.

Biotropic weather factors are assessed based on the degree of variability of meteorological parameters (Table 2.1). This classification has significant advantages because it takes into account the degree of expression of dynamic weather changes and the biometeorological effect on the body caused by a particular type of weather.

The classification is based on the assessment of oxygen concentration in the air as an integral indicator, the value of which is impacted by other meteorological parameters: air temperature, barometric pressure, and humidity. Weather changes can produce a variety of effects – hypoxic, spastic, hypotensive, stimulating (toning), and adiaphorous.

The hypoxic effect is caused by a decrease in the concentration of oxygen and is often observed in the zone of stable low pressure (a cyclone, a trough of low pressure, a low-pressure gradient) in the warm weather front. The most pronounced hypoxic effect is produced by a combination of hypobaric conditions and high humidity. In this case, a decrease in oxygen concentration in the air can be quite significant – from 300 to 270 g/m3.

The spastic effect is produced by an invasion of cold air (cold atmospheric front) and an increase in barometric pressure (ridge, low gradient high-pressure zone), often combined with the intensification of winds. Such weather conditions trigger angiospastic reactions in the body. The spastic effect can be caused by colder and stronger winds and stable hygrothermal conditions. The most pronounced spastic effect is brought about by a combination of hyperbaria, hypothermia, and high winds.

The weather causing hypoxic and spastic effects is often preceded by synoptic-meteorological conditions, which trigger the opposite reactions – those of hypotensive and restorative types. These weather conditions do not cause any adverse effects in the majority of patients. On the contrary, they reduce the risk of diseases of certain nosological groups. For example, the hypotensive type of weather is favorable for patients with hypertension. Patients with arterial hypotension benefit from the stimulating (toning) weather type. If the daily variations of all meteorological parameters are within the limits of their mean long-term values, the adiaphorous (neutral) type of weather is formed, which does not cause the exacerbation of any diseases.

Researchers from the Russian Centre for Restorative Medicine, Rehabilitation and Balneology (Moscow) hold that the spastic type of weather has a detrimental effect on the health of patients with hypertension. The highest incidence rate of acute myocardial infarction was observed during this weather, whereas the hypoxic type had a less damaging effect. Several lines of evidence suggest that in the arid zone of the Far South, a higher incidence rate of ambulance calls for ischemic heart disease was registered during the formation of the hygrothermal conditions characterized by considerable thermal discomfort leading to the development of hyperthermal weather hypoxia. In Saint Petersburg climate, spastic and stimulating weather conditions were the most unfavorable for patients with bronchial asthma.

During spells of hypoxic weather, patients with ischemic heart disease experience changes in microcirculation. In non-meteoliable patients, that is, those who are not vulnerable to weather, most changes were of adaptive character (an adequate increase of capillaries permeability; morphofunctional changes – the opening of "reserve" capillaries and the expansion of capillary network and an increase in the number of functioning capillaries). Meteoliable patients experience maladaptive changes – extensive capillary permeability, spasms of microvessels, especially arterioles).

Based on the 24-h Holter monitoring, meteopathic reactions in ischemic disease and hypertension patients during spastic weather conditions are characterized by the increased activity of the sympathetic nervous system, a higher vessel density, a lower cold tolerance, reduced physical exercise capacity, a higher incidence, and a longer duration of myocardial ischemia. Psychological tests revealed reactive anxiety, which increases in hypertension patients under the conditions of spastic and stimulating types of weather compared with the value of this indicator during continuously mild weather. Weather factors aggravated the meteosensitivity of hypertension patients, particularly after psychological stress.

During spastic weather, patients with bronchial asthma experience an activation of the sympathetic nervous system. The cyclical changes in the physical properties of air are determined by astronomical factors: the duration of day and night, a change of seasons, and the maximum height of the sun during a year depending on the geographical latitude of a territory [24]. Annual temperature variation, regardless of latitude, is expressed as a curve, with a decrease in temperature in January–February, and an increase in July–August. Daily air temperature amplitude is higher in the north and mid-latitudes, and lower in the south.

Weather changes are connected with aperiodic fluctuations of barometric pressure at the earth's surface. Daily fluctuations of barometric pressure in stable weather conditions are not more than 1–2 mm Hg. During more abrupt weather changes, barometric pressure fluctuations may exceed 10–20 mm Hg. Temperature distribution in the troposphere results in the formation of an area of low pressure in its lower layers at the equator, and an area of high pressure at the poles. In the high layers of the troposphere, the distribution of barometric pressure demonstrates the opposite trend. High-pressure air masses (anticyclones) usually bring fine weather. Anticyclones occur before each series of cyclones. In the periphery of cyclones, rain areas are usually formed by the influx of cold air.

The earth has a geomagnetic field, which extends outward. Rapid changes in the geomagnetic field, such as magnetic disturbances and magnetic (solar) storms, are caused by a stream of electrically charged particles from the surface of the Sun. The maximum frequency of magnetic storms has been registered during periods of intense solar activity, the so-called periods of solar turbulence. Atmospheric activity and electromagnetic oscillations caused by solar storms increase in summer and decrease in winter. Thus, the concept of *climate* includes not only temperature, humidity, movement of air masses, and changes in barometric pressure but also electromagnetic factors – the intensity of the magnetic field, air electric conductivity, electromagnetic oscillations, and solar radiation.

The quantity of solar thermal energy and the spectral composition of light depend on the angle of the sun. Biological rhythms are connected with the alternation of light and darkness, and their amplitude depends on the fluctuations of meteorological parameters: temperature, barometric pressure, humidity, wind force, etc. Consequently, the change of seasons, that is, the so-called "sunlight" climate, is one of the important factors to which the body reacts. Numerous clinical observations confirm that light is the main synchronizer of human biological rhythms. Diurnal rotation of the Earth causes wave-like changes in the functions of the body with a period close to 24 h. These rhythms are usually called circadian, and their frequency characteristics are different.

Professor Chizhevsky (1897–1964) repeatedly spoke about the existence of solar–terrestrial interactions. He proved that the intensity of electromagnetic radiation of the sun determines the intensity of biological processes in the body. He argued that during the period of active Sun, the number of epidemics increases. In 1930, Chizhevsky predicted the 1960–1962 epidemic of cholera in Southeastern Asia. He forecast nine influenza epidemics, eight of which did occur, including the one in Europe in 1968–1969 [25].

Numerous works by biologists, entomologists, and zoologists have shown that mass diseases of animals have a periodicity of about 11 years, that is, are synchronous with solar activity. Changes in solar activity and weather changes contribute to the worsening of chronic diseases in people.

Seasonal diseases and seasonal exacerbation of chronic diseases are the ultimate manifestations of the negative impact of climatic factors. Cold-related diseases (influenza, acute respiratory and inflammatory respiratory diseases, etc.) are most commonly associated with particular seasons of the year. The maximum incidence of these diseases is registered in autumn, winter, and early spring.

It is known that weather-related changes in the body are caused by a complex of factors, including not only the standard meteorological conditions (temperature, humidity, air mobility, barometric pressure) but also electrometeorological parameters (magnetic field of the Earth, air electric conductivity, etc.).

The history of climatology has documented days when a sudden change in weather led to the occurrence of many diseases. In January 1780, a sharp change of weather was registered in Saint Petersburg. During one night the air temperature rose from −44 °C to +6 °C. On November 18, 1954, in Tashkent, the air temperature suddenly dropped from +15 °C to −21 °C. As a result of such weather shocks, the incidence of colds increased and there were numerous cases of cardiovascular and bronchopulmonary pathologies. Adverse reactions of the body, caused by changes in weather conditions, are now called meteopathies.

Most diseases and their exacerbation are connected with sudden weather changes during the passage of a meteorological (weather) front – a transition zone between two air masses of different density and temperature. When the front is passing, all the meteorological parameters – barometric pressure, air temperature, humidity, wind speed, ionization, air conductivity, and natural radioactivity – change dramatically.

Changes in electrometeorological parameters during the passage of weather fronts have different stages. At first, only atmospheric electricity changes: electric field amplitude, air electric conductivity, and the intensity of electromagnetic impulses. During this period, other meteorological factors – air temperature and humidity, barometric pressure, and wind speed – do not exceed the ranges of their normal daily fluctuations. The second stage, the passage of a weather front, usually takes 1–6 h. This period is characterized by abrupt changes of all meteorological parameters, including atmospheric electricity. The third stage takes about a day, at which time meteorological and electrometeorological parameters return to their previous values.

The incidence of various pathological reactions and exacerbation of chronic diseases can be observed before, during, or after the passage of a weather front. Clinical observations indicate that the highest percentage of pathological reactions are registered 1–2 days before the passage of the front, that is, at the time of the most abrupt changes in atmospheric electricity. During this period, exacerbations occur in more than 70% of patients with hypertension, in more than 80% of patients with angina pectoris, and approximately in 70% of patients with eczema and pulmonary tuberculosis.

The correlation between the body's reaction to weather conditions and the type of higher nervous activity has been established. Patients with a stronger and balanced type of the higher nervous system react to the passage of a weather front by having only subjective sensations. Patients with a weaker and unbalanced type of the higher nervous system have objective symptoms of the worsening of their condition: a rise in arterial pressure, a change in the tonus of peripheral arteries, the lengthening of optical chronaxia, and changes in EEG. Moreover, there are also negative shifts in many biochemical processes – the level of sodium, cholesterol, and prothrombin in the blood increases, whereas the activity of some blood enzymes – catalase and peroxidase – decreases, etc.

Various types of climatic influence on the human body can significantly hamper adaptation as climate- and meteo-related pathological responses develop. Owing to a drastic climate change, climate-related pathological responses come in the form of cerebral, cardiac, vegetative-vascular, arthralgia, and other pathological symptoms; they also show metabolic disorders, depending on the particulars of the underlying and concurrent diseases and the characteristics of the unfamiliar climate. Meteopathic reactions most often develop in weather-sensitive people. Meteo-sensitivity can be of physiological and pathological types. The physiological type includes, for example, the meteorological sensitivity of the EEG alpha rhythm which is highly individual, variable, and, compared with other biological rhythms of the human body, is the most adaptive and sensitive of all bodily functions. It has been established that the maximum amplitude of the alpha rhythm changes following the change in wind direction. Such meteorological sensitivity is not associated with considerable shifts and disturbances in the metabolic energy of the brain; on the contrary, it rather exemplifies the subtlest changes within the normal range, limited by adaptive mechanisms, and with minimal energy losses. On the other hand, when weather-sensitive people move to the harsh climatic conditions of the Far North, especially in the winter season, they often develop pathological reactions, such as disturbances of higher nervous activity, respiratory and circulatory systems, thermal adaptation mechanisms, and changes in a mental state characterized primarily by anxiety and restlessness.

Meteo-related pathological reactions occur in the body in response to an abnormally sharp change in weather even in a familiar climate. On average, a person reacts (to varying degrees) to a temperature drop of 6 °C within 24 h, an atmospheric pressure drop of 3.75 mm Hg, and a decrease in the oxygen content in the air reaching 5 g/m3. These indicators characterize a human's threshold meteorological sensitivity. It has been shown that the occurrence of meteorological reactions reduces the effectiveness of treatment by almost 1.5 times. Therefore, an unfavorable weather forecast indicates that urgent preventive measures should be taken not only for individuals with enhanced weather sensitivity but also for those who show initial manifestations of meteorological responses. Meteo-related reactions are caused by maladaptation that occurs when the body is exposed to weather factors of above-threshold values. Individuals with psycho-emotional disorders show weather sensitivity in 80% of cases, and patients with musculoskeletal disorders in 60%. Negative meteotropic reactions develop, to a great extent, owing to the increased formation of free radicals, and a decrease in the antioxidant system activity. Unfortunately, an objective assessment of meteosensitivity presents certain difficulties. Meanwhile, in the context of global climate deterioration, the issues of diagnosing meteo-pathological reactions are of particular relevance. Thus, unfavorable weather and meteorological conditions can play a provocative role, triggering either latent pathological processes in weather-sensitive people or exacerbating their chronic diseases [26].

Most authors explain the mechanism of meteotropic reactions by electromagnetic impulses with the subsequent impact of meteorological factors (especially cold-weather ones) changing the body's reactivity to weather conditions in general. Electromagnetic pulses affect the functioning of the central nervous system, vascular tone, and metabolism, which exacerbates pathological processes and a subjective deterioration in well-being. Meteotropic disturbances develop in two ways. Firstly, unfavorable weather changes cause a complex of specific shifts in healthy people. In this case, meteorological factors trigger poor health symptoms caused by heliometeopathic or meteotropic reactions. The second way is the influence of weather changes on patients with pre-existing conditions. In addition, two other types of bodily reactions to weather factors are distinguished: on the one hand, meteotropic reactions associated with the inability of the body to maintain homeostasis, on the other hand, physiological adaptation to unusual climatic factors, that is, the process associated with the development of a new stereotype of internal reactions. It is generally accepted that the autonomic nervous system is the first to be affected when the weather changes. The weather factors change the tone of the sympathetic and parasympathetic divisions owing to the production of adrenergic and cholinergic mediators. The cause-and-effect relationships between meteotropic reactions and environmental conditions are still largely unclear. However, it has been established that meteotropic factors upset the assimilation/dissimilation balance in the body. Chemical processes are disrupted, and thermal, mechanical, and electrical energy is released. Electrical energy, which under normal conditions is formed in minimal quantities, is important in the development of meteotropic reactions as it influences the excitability of tissues, primarily the nervous system.

The development of meteotropic reactions is highly dependent on membrane permeability and excitation transfer, as well as disruption of intra- and extracellular metabolism. In addition, weather factors are thought to affect a decreased number, or transition of one type to another, of α-adrenergic receptors, H1- and H2-histamine receptors of the cell membrane. There are three phases in the development of meteotropic reactions:

1. The phase of clinical and physiological adaptation of the body to atmospheric and physical factors.
2. The phase of hypersensitivity to sudden weather changes, manifested by a change in mental and immunological reactivity.
3. The phase of maladaptation to the weather, expressed in healthy people by a functional weather-somatic syndrome, and in patients by the manifestation of sub-clinical and clinically pronounced reactions and aggravations.

Meteo-sensitivity is diagnosed by assessing the meteorological history and dynamic disease monitoring, against the background of weather and meteorological conditions. Meteoclimatic analysis implies the analysis of the clinical course of the disease, taking into account the time of exacerbation, seasonality, and the identification of meteotropic reactions to climate change in comparison with weather characteristics.

According to the degree of severity, three types of meteopathic reactions can be distinguished:

1. Mild reactions of the first degree, characterized mainly by subjective intoxication-free asthenic, artalgic, or myalgic syndromes, and most often expressed in one of them.
2. Moderate reactions of the second degree, with objective symptoms accompanied by intoxication, low-grade fever for 3–5 days, which does not affect the course

of the underlying chronic disease. These include mainly intercurrent diseases, mainly colds (upper respiratory catarrh, tonsillitis), vegetative dystonia syndrome, angiodystonic edema, hemorrhagic rashes.
3. Strongly expressed reactions of the third degree, with the exacerbation of the underlying disease (hypertensive crisis, angina attacks, exacerbation of chronic pneumonia, etc.). In such cases, various pathogenetic factors of cardiac, cerebral, or hypertonic type may occur. Thus, clinically meteopathic reactions of the third degree exacerbate the underlying disease (cardiological patients are the most meteolabile type, while patients with lung diseases, central nervous system, and musculoskeletal disorders are less susceptible to weather influences). The difference between meteopathic reactions and an exacerbation of the underlying disease consists in the fact that same-type patients develop simultaneous and massive pathological manifestations in adverse weather conditions; the short-term deterioration of their health is synchronized with weather changes; the same patient experiences relatively stereotyped repeated disorders in similar weather conditions [27].

There is a correlation between daily peak values of the electromagnetic field and the frequency of seeking emergency medical assistance by patients with cardiovascular diseases. The meteolability of men aged 50–59 years is the highest; among women, the age group of 18–29 years are the most sensitive to weather changes. It has been established that a meteotropic reaction in the critical age groups living in polluted territories occurs immediately on the day of geomagnetic disturbances, whereas in the presumably unpolluted territories it develops within 2 or 3 days. The same dependence has been observed in women aged 40–49 as a response to barometric pressure fluctuations.

Thus, the reasons contributing to the development of meteopathic reactions are very diverse. They include very rapid changes of weather and its components (aperiodic changes of barometric pressure, humidity, air temperature, wind, and precipitation), as well as the passage of weather fronts (cold, warm), the formation of cyclones and anticyclones (areas of low and high barometric pressure), and geomagnetic changes (magnetic storms, increase in solar activity, changes of electrometeorological conditions). All the enumerated factors contribute to the occurrence of pathological reactions in meteolabile people. This dependence should be taken into account in daily activities to prevent meteopathic reactions in chronically ill patients.

Chapter 3
Climatic Norms, Definition Periods: Methods for Determining the Areas of Biological Comfort/Discomfort

The unprecedentedly high rate of global warming and climate change in recent decades is a cause for concern. It has been established that human economic activity drastically affects climate. The diverse climate change effects manifest themselves in the frequency and intensity of weather and climate extremes. Expected climate change will inevitably affect people's lives in all regions of the planet, and in some of them it will become a tangible threat to the well-being of the population [28].

The average rate of climate warming for the land surface of the Northern Hemisphere is +0.328 °C for 10 years. On the territory of Russia, the rate of warming was +0.43 °C for 10 years. In all seasons, except winter, the warming rate slightly increased, and in winter, on the contrary, it noticeably decreased. Global warming, characterized by an increase in ambient temperature and humidity, affects human health, directly disrupting the body's thermoregulation with serious consequences: dehydration, fatigue, heat stroke, and even death [29].

Climate change is considered to be one of the leading factors influencing people's health. There is both a direct impact caused by an increase in the number of days with abnormally high and/or low temperatures, floods, storms, and typhoons and an indirect one brought about by environmental or socio-economic factors (an increase in the drylands area, a decrease in the volume of high-quality drinking water, etc.) [30, 31].

In Europe, WHO estimates that climate change causes between 1% and 10% of deaths in older age groups each year, and globally more than 150,000 additional deaths and 5.5 million years of disability per year. This represents 0.3% of total deaths and 0.4% of total years of disability, respectively. By 2050, a further increase in climate warming-related deaths of around 1–1.5% is expected. The economic costs of additional deaths due to climate change in the world vary widely, ranging from $6 billion to $88 billion a year [32].

1. The development of many regions of Russia is hampered by adverse environmental factors, in particular, heat or cold stress. According to the map "Territorial

A. V. Tarasov, R. S. Rakhmanov, *Marine Climate of Russian Coastal Territories*,
https://doi.org/10.1007/978-3-031-30951-9_3

zoning of the Russian North according to natural and climatic factors," such regions occupy 68% of the country's territory [33]. The author states that "Russia is a territory where cold stress conditions prevail. In winter, cold stress of various gradations prevails throughout the territory: extreme cold stress – in the north and northeast of the country from the north of the Arkhangelsk region to Chukotka; very strong cold stress – on the Kola Peninsula, in the Volga region, in the Urals, and the Asian part of the country. Most of the European territory and the south of Primorsky Krai experience severe cold stress, and moderate cold stress is observed in the south of the European territory. With climate warming at the beginning of the twenty-first century (2001–2010), cold stress weakens slightly. In summer, conditions range from moderate cold stress in the Arctic Islands and on the coast of the Arctic Ocean to 'no heat stress' in most parts of Russia and 'comfort' in the south of European parts."

Therefore, it is important

- To define the dynamics of multiyear climatic changes not only on a global scale but in specific territories.
- Using various bioclimatic indices, to assess climate comfort/discomfort of territories.

Meteorological indicators used to assess the impact of climate change on public health include average daily, minimum, and maximum atmospheric air temperatures, as well as radiation temperature; average daily relative humidity; atmospheric pressure; average daily, maximum, and minimum wind speed; and the amount of precipitation. The level of humidity in the air can have various negative effects on human health. At low humidity rates, dryness of the eyes, skin, and nasal mucosa is noted, which contributes to an increase in the incidence of colds [34, 35], especially in high air velocity conditions [36]. At high humidity rates and increased air temperature, thermoregulation can be disrupted, leading to such serious consequences as dehydration, fatigue, heat stroke, and even death [37, 38]. Labor productivity decreases and the risk of work injury increases [39]. It has been shown that higher relative humidity increases the risk of influenza breakout [40]. Ambient temperature may not always be a measure of the cooling and damaging effects on the body [41, 42]. Thermal sensations at the same temperature levels depend on wind strength and humidity. At the same time, high humidity can exacerbate the negative effects of wind. At low temperatures, meteorological sensations are exacerbated by both the strong wind and high humidity [43]. The role of humidity is especially significant at extremely low air temperatures, high wind speeds, and the loss of heat-shielding properties of clothing when it is moistened. In the absence of heating or dry clothes, health risks increase dramatically [44]. A higher wind velocity and lower humidity boost heat loss (a feeling of a decrease in air temperature), while a weakening of the wind and an increase in humidity leads to the opposite effect [45].

Individual human susceptibility to the effects of cold is characterized by great variability: diseases, gender, age, and constitutional features of the body can affect the frequency and severity of the cold exposure consequences. Based on the analysis of the literature and their own research, the authors provide a list of health effects caused by cold exposure. Cold can both be a causal (etiological) factor in the occurrence of health damage and act as a decisive trigger mechanism for other pathological processes (pathogenetic factor).

The authors have grouped diseases and other health effects whose occurrence, severity, and outcomes are etiologically and pathogenetically associated with exposure to cold.

Group 1: Diseases and health effects whose main cause or defining condition is cold exposure:

- Very high strength and stability of association with cold (acute lesions from exposure to excessively low natural temperature): hypothermia (acute); frostbite; urticaria caused by low-temperature exposure; chills; trench arm and foot; chronic lesions from exposure to extremely low natural temperature; cold injury syndrome; other specified polyneuropathies (polyneuropathy from exposure to excessively low natural temperature); rhabdomyolysis – severe damage to muscle tissue – (myopathy, myoglobinuria with renal failure due to hemolysis caused by low-temperature exposure).
- Diseases and other health damages in which exposure to cold is not the main cause but acts as a determining condition for the severity of their clinical manifestations: cold agglutinin disease; paroxysmal cold hemoglobinuria (Donath–Landsteiner syndrome); cryoglobulinemia; thromboangiitis obliterans (Buerger's disease); Raynaud's syndrome; other specified peripheral vascular diseases: acrocyanosis, vasomotor acroparesthesia (Nothnagel type); erythrocyanosis, erythromelalgia.

Group 2: Diseases and other health disorders, in whose occurrence the exacerbation of the clinical course and severity of cold exposure outcomes is a proven risk factor. High strength and stability of association with cold: chronic obstructive pulmonary disease; asthma; arthropathy; phlebitis and thrombophlebitis (superficial vessels of the lower extremities, femoral vein, other deep vessels of the lower limbs); skin-limited vasculitis, not elsewhere classified; dorsopathy.

Group 3: Diseases and other health disorders, in whose occurrence, exacerbation of the clinical course and severity of outcomes exposure to cold is a likely risk factor. The level of cold inducibility is probable: ischemic heart disease; heart rhythm disturbances; cerebrovascular disease; arteritis; intermittent claudication, arterial spasm; acute respiratory infections of the upper respiratory tract; other acute respiratory infections of the lower respiratory tract; spontaneous abortion, premature birth; edema, proteinuria and hypertensive disorders during pregnancy; acute and chronic prostatitis; acute and interstitial cystitis (chronic).

The results (consequences) of the dangerous effects of cold on the body are divided into five groups according to the degree of stability of the association and the evidence of its etiological or pathogenetic role in the occurrence, exacerbation of the clinical course, and worsening of the outcomes of diseases.

1. Indicators of physical condition: age; gender; low body weight; insufficient muscle mass; deficiency of body fat; consequences of cold injuries: numbness, erythema, local hyperhidrosis.
2. Behavioral risk factors for increasing the severity of consequences of dangerous exposure to cold:
    - Deviation, characterized by damage to adaptive behavior.
    - Mental and behavioral disorders associated with the use of psychoactive substances; behavioral syndromes associated with physiological disorders, including mental and behavioral disorders caused by the use of hallucinogens, cannabinoids, opioids, and cocaine; use of alcohol, volatile solvents, and tobacco.
    - Neurotic, stress-related and somatoform disorders that disrupt adaptive behavior in cold weather conditions.
    - Affective disorders (depression).
    - Lifestyle problems, including lack of physical activity.
3. Diseases accompanied by disruptions of thermoregulation, which can increase the body's susceptibility to the dangerous effects of cold: polyneuropathy, dysfunction of the hypothalamus; Parkinson's disease; multiple sclerosis; clinically expressed forms of malignant neoplasms; acute and chronic renal failure; acute infectious diseases, diabetes mellitus; hypotension; thermal and chemical burns; psoriasis; acquired and congenital ichthyosis; dermatitis and eczema; diseases of the neuromuscular synapse and muscles; Raynaud's syndrome; carpal tunnel syndrome, anemia; systemic lupus erythematosus.
4. Functional states of organs and systems that reduce the effectiveness of body thermoregulation:
    - Hypokinesia and hypodynamia, gait disturbance.
    - Overtired state. Chronic fatigue syndrome, malaise and fatigue.
    - Hypothyroidism.
    - Hypoglycemia.
    - Hypocorticism (chronic insufficiency of the adrenal cortex).
    - Hypoglobulinemia.
    - Blood clotting disorders: respiratory distress syndrome (DIC) 1–2 stages; other clotting disorders.
    - Hyperhidrosis.
    - Systemic disorders of microcirculation and capillary blood flow.
    - Immobilization syndrome, limitation of limb mobility.

Group 5: The consequences of the adverse effects of drugs, medicines, and biological substances used for therapeutic purposes:

- Adrenolytic drugs: alpha- and beta-adrenoblockers (they boost heat transfer owing to increased blood flow in peripheral vessels).
- Anticholinergics: n-anticholinergics (muscle relaxants and ganglion blockers) m-, n-cholinomimetics, incl. Anticholinesterase agents (they increase heat transfer

as a result of the expansion of peripheral vessels; reduce the effectiveness of hormonal regulation of heat transfer and contractile thermogenesis; affect temperature sensitivity as a result of inhibition of the transmission of impulses from cold receptors).

- Neurotropic drugs: anxiolytics (they reduce the efficiency of heat generation as a result of a decrease in synaptic reflexes and inhibition of the activity of the hypothalamus; disrupt adaptive behavior in the cold due to a hypnotic effect).
- Myotropic antispasmodics (reduce the effectiveness of contractile thermogenesis).
- Genitourinary system regulating drugs, and diuretics (they increase heat loss with excreted urine and increase blood viscosity as a result of dehydration).
- Adrenal cortical, hypothalamic, and thyroid hormone antagonists (they affect the hormonal regulation of heat production and dissipation).
- Ciprofloxacin (increases heat loss as a result of hyperhidrosis).
- Propranolol (heat production disorders).
- Sedatives and hypnotics (behavioral disorders).

3.1 Periods for Determining Climate Normals

The weather and climate characteristics of the territories are defined on the basis of long-term observations. According to the World Meteorological Organization (WMO), climate normal periods are based on 30 years of data: the first period established by WMO was 1931–1960; the second was 1961–1990; and the third was 1991–2020. Since 2021, the fourth period of establishing climate norms has begun [46].

In order to determine changes in climatic conditions in certain areas, a study was conducted on subarctic and continental (temperate) climates. The data was obtained from the Meteo.ru website.

We analyzed the average monthly indicators in the open air: ambient air temperature (°C), air velocity (wind, m/s), and relative humidity (%) determined in two periods of climatic norms: 1961–1990 and 1991–2020. The duration of the warm and cold periods of the year was determined; to that end, we used data from each period's last decades – 1981–1990 and 2010–2019 accordingly.

The analysis of atmospheric air temperature in each observation period shows that the warm period of the year in a temperate climate lasted 5 months (summer, May to September) in the first period, with temperatures ranging from 9.2 °C to 18.1 °C and 8.0 °C to 10.4 °C; the second period also lasted 5 months (9.4 °C–18.8 °C). The cold period took up the remaining 7 months of the year. In April and June, the temperature in the last observation period was significantly higher than in the previous one, respectively, by 4.0–3,9 °C (Fig. 3.1); in July–October it was higher by 0.7–0.8 °C.

In the subarctic climate, the warm span of the year in the first observation period lasted only 3–4 months (summer period, September as a transition from warm to

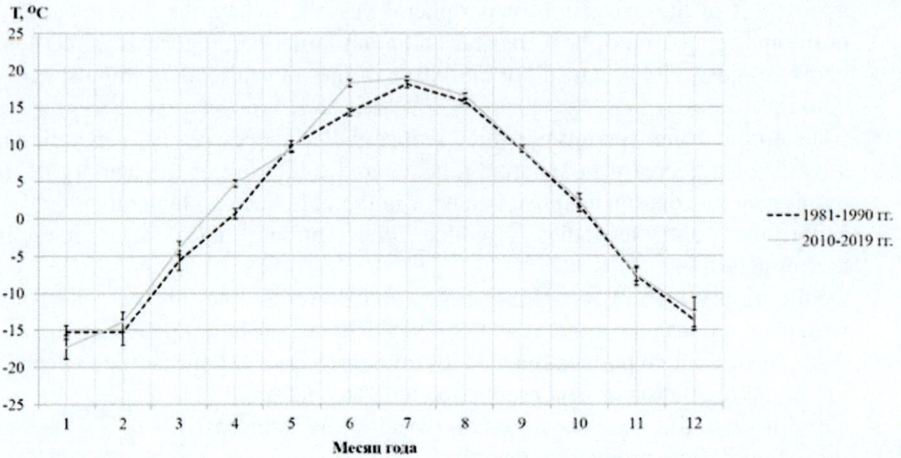

Fig. 3.1 Indicators of atmospheric air temperature in a temperate continental climate during climatic normal periods

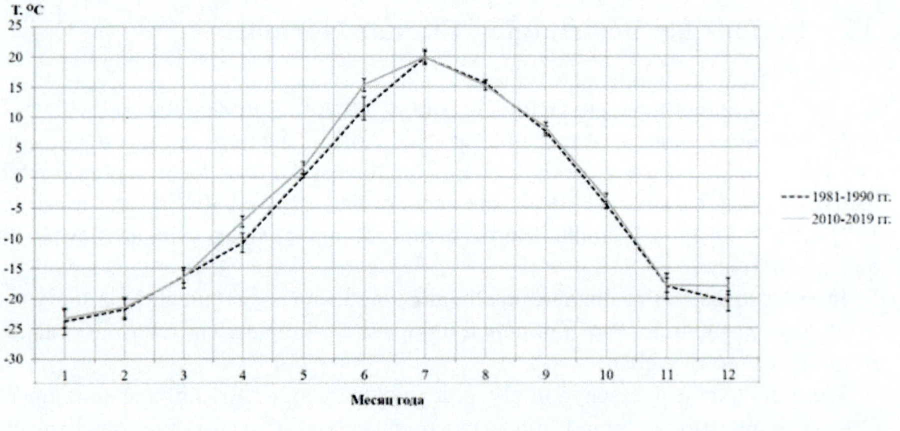

Fig. 3.2 Indicators of atmospheric air temperature in the subarctic during climatic normal periods

cold); the cold period covered 8–9 months. According to the second observation period, the warm period of the year also lasted 3–4 months, with the rest of the year staying cold (Fig. 3.2). The temperature was statistically significantly higher only in April ($p = 0.006$).

In the temperate climate region, changes in humidity were determined by observation periods. In September–May, relative humidity in the second observation period was higher by 1.7–6.6% compared to the previous observation; in November–March, the differences were significant: November – $p = 0.012$; December – $p = 0.023$; January – $p = 0.05$; February – $p = 0.05$; and March – $p = 0.042$ (Fig. 3.3).

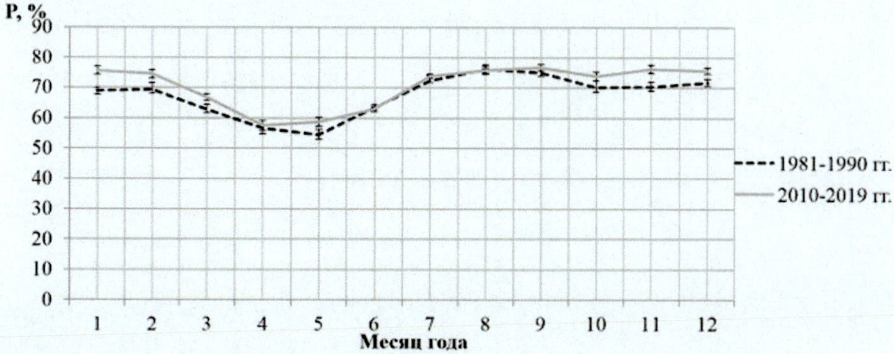

Fig. 3.3 Indicators of relative air humidity in a temperate climate during climatic norm periods

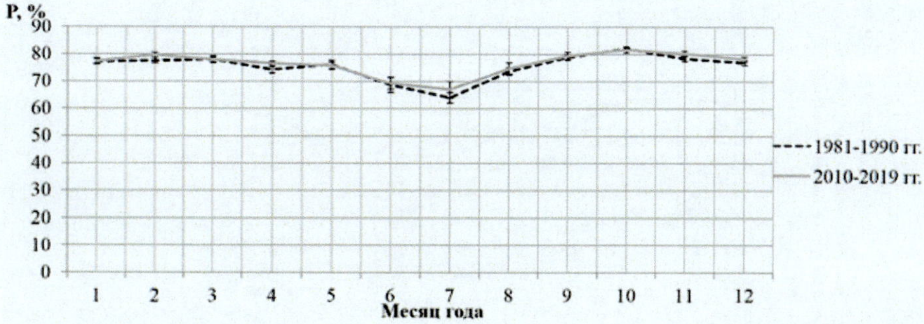

Fig. 3.4 Indicators of relative air humidity in the subarctic during climatic normal periods

In the subarctic, no changes in relative humidity indicators were found over the observation periods (Fig. 3.4). At the same time, rather high humidity was noted in the autumn, winter, and spring periods.

Under the conditions of a temperate continental climate, the second observation period showed a significant decrease in wind speed in all months, except for September (Fig. 3.5). Wind speed dropped from 0.2 m/s (August, $p = 0.038$) and 0.4 m/s (February, $p = 0.045$) to 1.0 m/s (November, $p = 0.001$). In the remaining months, a significant decrease in wind speed was 0.6 m/s.

Rather high wind speeds were noticeable in the subarctic climate (Fig. 3.6), with average monthly values reaching 6.1 m/s and above in the first observation period, and 5 m/s in the second one. Wind speed decreased significantly by the month: in April–December, a noticeable drop of 0.7 m/s in wind speed was observed (August, $p = 0.015$) to 1.1 m/s (November, $p = 0.041$). In February–March, the wind speed was also lower by 1.0 m/s, respectively, but these changes were not significant ($p = 0.09$ and $p = 0.71$). Only in January the wind speed did not change in the second observation period.

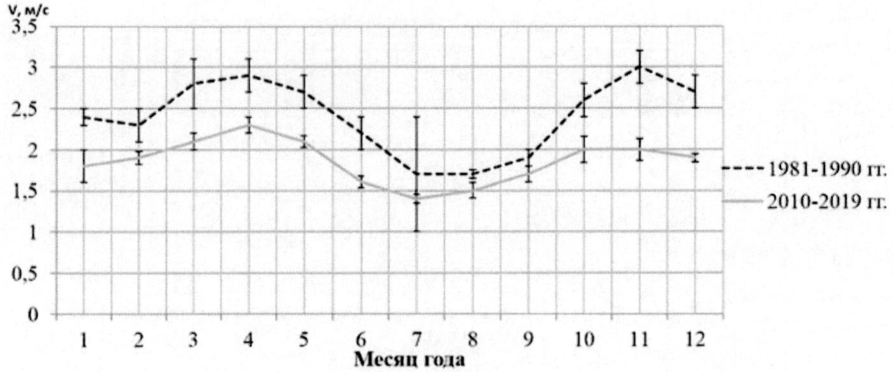

Fig. 3.5 Wind speed indicators in a temperate climate during climatic normal periods

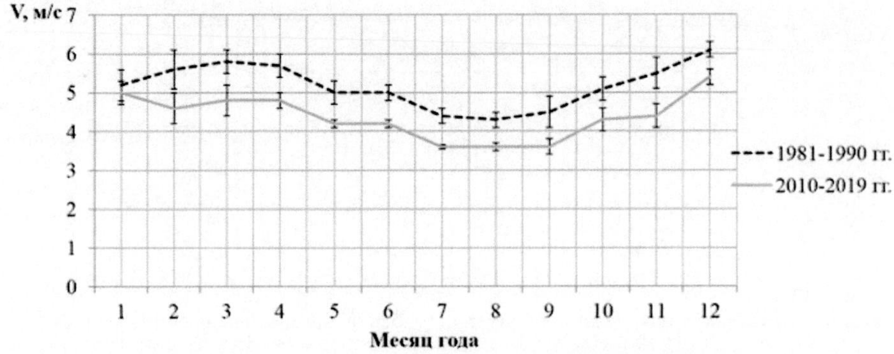

Fig. 3.6 Wind speed indicators in the subarctic during climatic normal periods

Thus, in the third period of establishing the climatic normals, all three estimated physical parameters of the external environment changed: atmospheric temperature, wind velocity, and humidity.

In a temperate continental climate, a significant increase in air temperature was noted in April–June, and in July–October the growth reached 0.8 °C. Air humidification occurred in September–May, with relative humidity going up to 6.6% ($p = 0.001$). Eleven months of the year showed a significantly lower wind speed than in the compared previous period.

In a subarctic climate, temperature values changed data, but only in the April–June period. The humidity did not show much change, remaining elevated for 10 months. The wind speed, as in a temperate climate, decreased: except for 1 month, the drop reached 1.1 m/s. The study proves the ongoing changes in the intensity of physical factors in each territory, but it also determined the features of these changes. Climate warming being the general trend, its patterns are diverse. Apparently, the changes will affect the ecological situation and human beings in different ways, which should be taken into account in medical and social forecasting and in planning work in the open.

3.2 Bioclimatic Indices

Currently, several indices are used to assess the bioclimatic comfort of the territory and health risk [47, 48]. Each physical environmental factor has its biological significance. To assess thermal comfort/discomfort in specific weather and climatic conditions, various complex indices are used alongside single-factor indicators – for example, when determining the seasonality index or the influence of heat or cold waves. However, the informative value of these indices may differ. For example, in cold weather conditions, air temperature and wind speed (cold indices) are the main (significant) safety factors (or frostbite risks) for working in an open area – an integral indicator of body cooling conditions (IIBCC), and Siple's wind-chill index (WCI); in hot conditions – temperature, humidity, and wind speed (effective temperature according to Steadman (ET)). IIBCC and WCI show the potential effect of cold at negative temperatures, while ET indicates that there is no health threat for a clothed person in the temperature range from −28 to 27 °C. The limitation of cold stress indices is that in calculations and experiments the human body is usually replaced by a dummy, which makes it impossible to take the effect of radiation heating into account.

According to Missenard, the effective temperature index (ET) makes it possible to determine thermal comfort/discomfort, that is, the effective temperature felt by the human organism in all seasons of the year.

To determine the cold risk, the most preferable index is the integral indicator of the body cooling conditions (IIBCC). It is used to devise the labor regime for working in an open area. The WCI is used to identify hazards to a working person under the influence of physical factors in a cold environment or indoors in order to identify problems and establish management methods aimed at eliminating or reducing sources of health hazards.

To assess the microclimate of a cold environment, the WCI is used in the form of an equation for the cooling effect of the environment (Siple, Passel, 1945). In recent years, the Universal Thermal Climate Index (UTCI) characterizing thermal comfort for a person has been increasingly used to assess the health risk of weather and climatic conditions. It allows a comprehensive assessment of the impact of such physical factors as wind speed, temperature, humidity, and radiation temperature.

3.2.1 The Integral Indicator of Body Cooling Conditions (IIBCC)

The IIBCC allows you to distinguish several degrees of risk of frostbite in the unclothed areas of the human body: no risk (value ≤34), the duration of a safe stay in the cold is not limited; moderate risk (˂34–≤47), the duration of stay is under 60 min; critical (˂47–≤57), the duration of stay is limited to 1 min; catastrophic (>57), the duration of stay under 0.5 min.

The IIBCC is determined according to the equation:

$$\text{IIBCC} = 34.654 - 0.4664 \times Ta + 0.6337 \times V$$

where

Ta – air temperature, °C.
V – wind speed, m/s.

The Wind Chill Index (WCI) corresponds to the air temperature in an open area, which, at a wind speed of 4.2 km/h, causes the same cooling effect as the actual environmental conditions. It characterizes the degree of wind cooling effect – the equivalent air temperature equal to the air temperature having a cooling effect in the absence of wind in the shade, without taking evaporation into account. More precisely, it is not a temperature parameter, but an index that helps to relate the cooling effect of the wind to the air temperature in calm conditions. Wind does not make an open object colder than the surrounding air – a higher wind velocity will only lead to cooling the object to ambient temperature in a shorter time. The cooling rate is calculated using a weighted average skin temperature of 33 °C. With complete calm and relative humidity of 100%, a person's thermal sensations depend only on the ambient temperature. At the same temperature, but with increasing wind and decreasing humidity, heat loss increases, and a person feels as if there was a decrease in air temperature. The opposite effect occurs when the wind weakens and humidity increases.

$$\text{WCI} = 33 + \left(0.478 + 0.237 \times \sqrt{v} - 0.0124 \times v\right) \times \left(T - 33\right)$$

The criteria for determining the risk of hypothermia in exposed parts of the human body are as follows:

- 10 ≤ −24 °C – discomfort, coolness
- 25.0 to ≤ −34.0 °C – very cold, hypothermia of the skin surface
- 35.0 to ≤ −59.0 °C – extremely cold, unclothed parts of the body can become supercooled in 10 min.

From −60.0 °C – extremely cold, unclothed parts of the body can get supercooled in 2 min [49].

3.2.2 Cold Wind Index (CWI)

$$\text{CWI is determined by the formula : CWI} = \left(10\sqrt{V} + 10.45 - V\right) \times (33 - t)$$

where

v – wind speed (m/s),
t – temperature (°C).

As shown by Morris, a human dressed in winter clothes, with thermal insulation from 0.451 to 0.528 °C m/W, in compliance with regulatory requirements, is supercooled at CWI = 1190.0–1547.0 kcal/m^2/h. Comfortable conditions correspond to the value of CWI = 761.6 kcal/m^2/h.

The effective temperature is a biometeorological index that characterizes the effect of a complex of meteorological elements (temperature, air humidity, and wind) on a person through a single indicator – the so-called effective air temperature.

A group of American scientists conducted a comparative analysis of several widely used algorithms for calculating the effective air temperature and came to the conclusion that the algorithm developed by Steadman is the most complete. To develop this model, a wide range of biometric measurements were used, taken in many countries from 1940 to 1995. The effective temperature model combines the body and skin physiological factors, the physical characteristics of clothing and the air layer in close proximity to the body, as well as meteorological environmental factors.

The body's resistance to the environment depends on the physical attributes of the individual. Therefore, the model is designed for the typical adult human of either sex, of average build, dressed for the weather, and walking in the shade at 4.8 km/h. Based on this model, Steadman derived simple formulas for calculating the effective temperature. Within a 95% confidence interval, their error does not exceed 1 degree Kelvin.

$$\text{Tef} = -2.7 + 1.04T + 2.0P - 0.65v,$$

where

T – air temperature (°C)
P – partial pressure of water vapor (kPa)
v is the wind speed at 10 m above ground level.

The effective temperature combines two previously used indices: wind-adjusted air temperature (wind chill) and humidity-adjusted air temperature (heat index). Negative values of the effective temperature indicate the probability of frostbite, while positive values indicate the probability of heatstroke. A range of −28 to 27 °C is defined as safe for a clothed person (Table 3.1).

Table 3.1 The effective temperature scale

˂ −50 °C	Dangerous – frostbite of exposed skin is possible in less than 5 min
−38 to −50 °C	Be extremely careful – frostbite of exposed skin areas is possible after 10–15 min
−28 to −38 °C	Be careful – frostbite of exposed skin areas is possible in 20–30 min
−28 to 27 °C	There is no danger for a person dressed for the weather
27 to 32 °C	Be careful – Fatigue is possible during prolonged active loads in the open air
32 to 40 °C	Be extremely careful – Sunstroke is possible during prolonged activities in the open air
40 to 55 °C	Dangerous – a high probability of sunstroke and overheating, heat stroke is also possible
>55 °C	Extremely dangerous – Rapid heat – Or sunstroke may occur

The effective temperature is determined by the Missenard formula (ET, °C). The ET index combines physical factors that produce the same thermal effect as still air at 100% relative humidity and a certain temperature; it estimates the thermal sensations of a human stripped to the waist.

$$\mathrm{ET} = 37 - \frac{37 - t}{0.68 - 0.0014 \times f + \dfrac{1}{1.76 + 1.4 + v^{0.75}}} - 0.29 \times t \times \left(1 - \frac{f}{100}\right)$$

- t – temperature, °C,
- f – relative air humidity, %,
- Interpretation of ET, °C is carried out according to the following criteria:

˂−24 (threat of frostbite)
18–≤ −24 (very cold)
12–≤ −18 (cold)
6–≤ −2 (moderately cold)
6–≤0 (very cool)
0–≤6.0 (moderately cool)
6.0–≤12.0 (cool)
12.0–≤18.0 (comfort, moderately warm)
18.0–≤24 (comfort-warm)
24.0–≤30.0 (moderate heat load)
>30.0 (heavy heat load).

To assess working conditions in the warm period of the year, the HLI (environment heat load index) is devised with four climate variables: air temperature, humidity, speed, and heat exposure.

The HLI index (WBGT index – psychrometer wet-ball temperature index) is an empirical integral indicator (expressed in °C), reflecting the combined effect of air temperature, air velocity, humidity, and thermal exposure on the heat exchange between a human and the environment.

When taking measurements outdoors, with solar heat load (or indoors under thermal radiation), the HLI is derived from the following formula:

$$\text{WBGT} = 0.7T_{\text{w}} + 0.2T_{\text{g}} + 0.1T_{\text{d}}$$

where

- T_{w} = natural wet-bulb temperature (combined with dry-bulb temperature indicates humidity), °C.
- T_{g} = globe thermometer temperature (measured with a globe thermometer, also known as a *black globe thermometer*), °C.
- T_{d} = dry-bulb temperature (actual air temperature), °C.

Indoors, or when solar radiation is negligible, the following formula is often used:

$$\text{WBGT} = 0.7t_{\text{w}} + 0.3t_{\text{g}}$$

Globe thermometer temperature (the temperature inside the blackened ball) is measured by a thermometer that is placed in the center of the hollow blackened balloon; Tg reflects the effect of air temperature, surface temperature, and air velocity. The blackened sphere shall have a diameter of 90 mm, the smallest possible thickness, and an absorption coefficient of 0.95. The temperature measurement error inside the ball shall be no more than ±0.5 °C.

The HLI is recommended for the integral estimation of environment heat load at workplaces where the air velocity does not exceed 0.6 m/s, and the intensity of thermal radiation is 1200 W/m2.

In accordance with GOST R ISO 7243-2007, if the environmental parameters do not have a constant value in space, it is recommended to determine the index in three positions corresponding to the height of the head, abdomen, and ankles relative to the ground. If the worker is standing, measurements should be taken at 0.1, 1.1, or 1.7 m above the floor, if they are sitting 0.1, 0.6, or 1.1 m above the floor.

If the analysis performed prior to the thermal overheating at the point under study showed that the environment was in fact homogeneous (<5% heterogeneity), a simplified procedure may be applied, consisting in determining a single index at the abdominal level only.

The UTCI can be applied in the key areas of human biometeorology, daily forecasting, regional planning, and environmental epidemiology, and it also works for all climate zones.

The UTCI index is expressed as the equivalent ambient temperature (°C) that has the same physiological effect on humans as the actual environment. The index has been extensively tested in various climatic conditions from Svalbard to Madagascar and has shown good sensitivity to environmental changes, namely, temperature, solar radiation, humidity, and especially wind speed. The index shows the human body's response to changes in these parameters. Comparison of the UTCI index with other known bioclimatic indices shows that it correlates well with indices that were obtained from various human thermal balance models and worse with indices based on relatively simple two-parameter formulas. One possible reason for the discrepancy is the absence of the radiation factor in the equations. Unlike other bioclimatic indices that are closely related to temperature, the UTCI index uses a more modern model of clothing insulation. The UTCI index represents bioclimatic conditions in terms that apply to humans over a wide range of climatic conditions.

UTCI can be applied in key areas of human biometeorology, daily forecasting, regional planning, and environmental epidemiology; it is also applicable in all climatic zones.

Initiated by the Commission of the International Society for Biometeorology and developed with the support of the European Union, this index aims to assess thermal environmental conditions. It is based on the Fiala multinode model of human heat balance. UTCI-Fiala mathematical model of human temperature regulation forms the basis of the new Universal Thermal Climate Index (UTCI). The UTCI index can be described as the equivalent ambient temperature (°C) that has the same physiological effect on humans as the actual environment. The offset or deviation of UTCI from air temperature (Ta) depends on the actual values of air temperature, mean radiant temperature (Tmrt), wind speed (va), and humidity expressed as water vapor pressure (e) or relative humidity (R):

$$\mathrm{UTCI} = f\left(\mathrm{Ta;Tmrt;va;e}\right) = \mathrm{Ta} + \Delta\left(\mathrm{Ta;Tmrt;va;e}\right).$$

In developing the index, meteorological and non-meteorological parameters (metabolic rate and thermal insulation of clothing) for standard conditions were determined as follows:

- Wind speed (v) 0.5 m/s at an altitude of 10 m (approximately 0.3 m/s per 1.1 m).
- Average radiation temperature equal to the air temperature.
- Water vapor pressure (e) corresponding to a relative humidity of 50%, and constant at high air temperatures (>29 °C), equal to 20 hPa.
- The activity of a person walking at a speed of 4 km/h (or approximately 1.1 m/s). This provides a metabolic rate of 135 W/m2.

The sensitivity of UTCI to temperature, humidity, radiation, and wind speed shows that it is applicable in warm and cold conditions. UTCI is classified in terms of human thermal exposure as follows:

- Above +46 °C – extreme heat stress.
- From +38 to +46 °C – very high heat stress.
- From +32 to +38 °C – severe heat stress.
- From +26 to +32 °C – moderate heat stress.
- From +18 to +26 °C – comfort.
- From +9 to +18 °C – no heat stress.
- From 0 to +9 °C – mild cold stress.
- From −13 to 0 °C – moderate cold stress.
- From −27 to −13 °C – severe cold stress.

- From −40 to −27 °C – very strong cold stress.
- Below −40 °C – extreme cold stress.

3.3 Problems of Acclimatization and Adaptation

Acclimatization is defined as a process of adaptation of biological species to life in new climatic and geographical conditions [50]. The acclimatization of animals and plants is usually interpreted as the interaction of two systems: biological species and the new environment they have to adapt. Not only unusual climatic and geographical factors but also living conditions are important factors for acclimatization. Comfortable living conditions and clothes, well-suited for a given climate, rational work-rest schedule, proper nutrition, a high level of welfare, and qualified medical assistance contribute to the adaptation of people to new and often harsh climatic and geographical conditions.

In this context, the acclimatization of people is partly a social process since the geographical environment affects people not only directly but also indirectly through the social conditions they live in. Living conditions play an exceptionally important role in the lessening of the negative environmental impact on the body. Long-term observations of acclimatization of people who have moved to the Far North and the southernmost regions of Russia show that the main factor influencing the process of acclimatization is not the severe climatic and geographical environment but rather favorable conditions of everyday life and work. Acclimatization has legal, social, ecological, as well as health and hygiene aspects, which are of the utmost importance and should be given a high priority.

Acclimatization is a long-term adaptation to new climatic and geographical conditions, which is also connected with the formation of a new dynamic stereotype produced as a result of establishing temporary and permanent reflexive connections with the environment through the central nervous system.

Adaptation is undoubtedly one of the fundamental characteristics of living matter. It is inherent to all known forms of life and is so universal that it is often identified with the very concept of life.

"Each living organism is a complex isolated system, the internal forces of which during every moment of its existence are in a state of balance with the external forces of the environment"(Pavlov).

Adaptation is usually studied in two aspects – static and dynamic. The static aspect reflects the properties (state) of a biosystem, its ability to resist the impact of environmental conditions. This ability is often analyzed in the context of the sustainability of a biosystem, which means that the biosystem can preserve its normal vital functions while being influenced by various environmental factors. The dynamic aspect of adaptation reflects the process of adaptation of a biosystem to the constantly changing environment. Biosystems have to change over time to maintain their vital functions irrespective of any changes in the external environment. The analysis of the dynamic aspect of adaptation is a study of adaptation mechanisms, processes, characteristics, the principles of regulation, etc.

The mechanisms of adaptation of the biological system to "adequate" environmental conditions are the result of a long evolution and ontogenesis. Adequate conditions of the environment are those that correspond to the genetic and phenotypic constitutional properties of the body at the given moment of its existence. "Inadequate" environmental conditions are those that do not correspond to the genetic and phenotypic properties of a biosystem at the given moment [51].

The functioning of a living organism in inadequate environmental conditions requires the activation of additional mechanisms. It has been suggested that adaptation is the functioning of a living organism under inadequate environmental conditions while still maintaining the optimal ratio of vital functions, and the ability to work and study. Insufficiency of compensatory adaptive mechanisms or their disruption results in the pathology of adaptation.

Thus, depending on external conditions (which can be "adequate" or "inadequate" for a biosystem), several qualitatively different states of the body can be distinguished: physiologically normal state, stress state, adaptation state, and pathological state.

Acclimatization occurs if external conditions do not necessitate going beyond the normal functions of the body and no compensatory mechanisms are required. Acclimatization is a physiological phenomenon, which can be defined as the ability of the body to successfully cope with new climatic and geographical conditions. Provided new climatic conditions require exceeding the body normal capabilities, a state of decompensation occurs causing specific pathologic reactions.

Adaptation triggers different physiological processes as a response to a particular type of climate. For instance, during acclimatization to a hot climate, there are changes in the normal functioning of the cardiovascular system (heart rate decreases and blood pressure rises by 15–25 mm Hg); respiratory system (respiratory rate decreases); excretory system (more intensive and even sweat evaporation and reduced stress on kidneys), which results in a 10–15% decrease in the main metabolism and body temperature. Acclimatization to cold, severe, and polar climate results in increased metabolism and thermogenesis (metabolic heat production), and higher circulating blood levels, etc. Previous studies have established that acclimatization to a hot climate is more difficult than to a cold one. The hot climate is typical of deserts and semi-deserts where air temperatures remain high (50 °C or more) for 5–7 months. There are sharp fluctuations of air temperatures from extremely high during the day and very low, down to −10 °C at night. Intense solar radiation, high temperature of all surrounding objects and soil, low relative humidity (12–20%), and dust storms complete the picture of a hot and dry climate.

Almost constant high air temperature (above 30 °C) during the whole year and high relative humidity hamper the elimination of body heat. The only processes that help maintain the thermal balance are the evaporation of sweat and the release of heat with breathing. The heat balance is quickly disturbed which negatively affects work performance and basic metabolism. The main reaction to heat is the expansion of peripheral blood vessels, which leads to a significant increase in the circulating blood volume and a decrease in arterial pressure. This reduces the functional capabilities of the cardiovascular system. Blood vessels of internal organs (liver, kidneys, and intestines) narrow to maintain the adequate skin blood flow. Intense sweating eventually leads to dehydration and blood thickening. The body loses vital water-soluble vitamins and salts with sweat.

When the bodyweight drops more than 15% due to excessive sweating, irreversible changes occur in the cardiovascular and nervous systems. Hot winds with dust damage the mucous membranes of the upper respiratory tract, nasal turbinate bones thicken, breathing through the nose becomes difficult, and the incidence rate of acute and chronic rhinitis and pharyngolaryngitis increases. Nasal filtration capacity and bactericidal properties of mucous membranes worsen, resulting in bronchitis and damage to the pulmonary parenchyma. Intense evaporation of sweat, exhausting heat, and the need to quench constant thirst cause disorders of the water-electrolyte balance and lead to the development of heat exhaustion.

It has been argued that a hot climate has a negative effect on digestion, which is manifested in reduced salivation, weaker tone and motility of the gastrointestinal tract, and higher acidity of gastric juice. This leads to a higher incidence of hypacid gastritis among those who have relocated to a southern region [26].

Cold climate is typical of subarctic (subantarctic) and arctic (antarctic) zones. In the subarctic (subantarctic) zone, the climate is mainly continental subarctic and subantarctic. Winters are long and severe. The average temperature of the warmest month is no higher than 12 °C, and the annual precipitation is less than 300 mm. The Arctic Ocean fuels strong cyclones bringing windy, cloudy weather and heavy precipitation.

The Arctic or Antarctic climate is characterized by harsh long winters, cool and short summers, and low annual precipitation (100–300 mm). The prevailing types of landscape are tundra and ice. The Arctic regions of Russia include areas of the Far North located to the north of the Arctic Circle. In some places, the temperature sometimes drops to – 40–50 °C in winter. Strong winds often blow at a speed of 20–30 m/s in blizzards and snowstorms. The cold season lasts 6–10 months. In spring and summer, the snow cover intensively reflects sunlight.

One of the main characteristics of the Arctic climate is a specific light condition – the polar day and the polar night, which affect all types of human activity in the Far North. Russian regions located in the North have the following climatic characteristics: low ambient temperatures, frequent fluctuations of the geomagnetic field, changes in the partial pressure of oxygen, sharp fluctuations in barometric pressure, lack of solar radiation, changes in air humidity, strong winds, the dependence of natural light conditions on the time of year (polar day and polar night), rapid change of meteorological conditions, surface inversions, ice spikes, poor visibility, frequent auroras in autumn and winter that contribute to the emergence of illusory sensations.

Another characteristic feature of polar winter in Zapolyarye is a sharp variability of air temperature. During the passage of air fronts, aperiodic air temperature fluctuations are up to 15–20 °C per day. This means that special winter and "arctic" clothes are needed to safely protect people from cooling for a long time. Particularly common are cold-related damage to the distal parts of the extremities, feet and hands: frostbites, chilblains, neurovasculitis, which may eventually progress to gangrene. The so-called trench foot is also widespread. It used to be quite common among soldiers, fishermen, and woodcutters. The lesion is the result of exposure to damp freezing temperatures from 0 to −10 °C during the first 10–12 h.

Seasonal factors in regions with a cold climate may cause snow ophthalmia, sleep disorders, and increased irritability. Insufficient intake of fresh food, especially fruits and vegetables, leads to hypovitaminosis. These conditions impose additional stress on the body and can contribute to increased cold-related morbidity and the exacerbation of chronic diseases [4].

Chapter 4
Adaptation to Marine Climate

The adaptation reactions of people living in the geographical conditions of the monsoon climate, particularly those living on the coasts, are of considerable importance. Monsoons soften the effects of high temperatures. At the same time, the rainy season complicates normal daily activities and triggers adaptive reactions to extremely high humidity and relatively high air temperature. The impairment of thermoregulation due to reduced cooling from sweating, especially during intense physical activity, contributes to hyperthermia and the overheating of the body. As a result, in hot and humid monsoon weather, the pressure on the cardiovascular system and other regulatory systems increases.

The extratropical monsoon climate does not have extremely hot weather. However, people living in temperate latitudes are exposed to rather high air temperatures in summer and low temperatures in winter due to the high seasonal circulation of air masses. During winter monsoons, the body metabolism increases, the body temperature and oxygen intake are slightly higher, the sympathetic nervous system and blood vessels tone are higher, and blood pressure and hematopoiesis also increase. There is a rise in the level of erythrocytes, leukocytes, neutrophils, and monocytes. During summer monsoons, the basic metabolism, body temperature, oxygen intake, blood vessel tone and blood pressure decrease, the parasympathetic system tone is clearly increased, the number of erythrocytes and their osmotic resistance slightly decrease, and there are often signs of leukopenia.

Data from several studies suggest that there are considerable seasonal variations in the process of adaptation. These variations contribute to the impact of temperate monsoon climate on the body and influence the incidence rate of certain diseases [52]. Geographic areas having a strong seasonal circulation of air masses in summer and autumn have a lot of sunny days similar to those in the Mediterranean. People moving from different inland regions to the coastal zone of the Mediterranean and Black Sea type still go through a period of acclimatization, which is characterized by a number of adaptive reactions. These reactions may be general in character as a

A. V. Tarasov, R. S. Rakhmanov, *Marine Climate of Russian Coastal Territories*,
https://doi.org/10.1007/978-3-031-30951-9_4

general response to new climatic conditions and specific as a response to a specific feature of this or that climate.

One of such specific characteristics of the maritime climate is the chemical composition of the air. Considering that an adult at rest breathes in about 10 m^3 of air a day, it becomes obvious that inhaling tiny droplets of seawater can have a positive effect on the adaption to the marine climate of all geographic zones. Long stay on the coast or near it, especially in stormy weather and breathing in diffused sea water lead to some biochemical and physiological changes, which are the most evident during the first days and weeks after relocation. During this period, water-salt metabolism increases, and diuresis and water retention also increase within physiological norms. The body gets vitally important iodine and bromine. It has been demonstrated that adaptation intensifies calcium phosphate exchange and enzymatic processes [53]. The functioning of the vegetative nervous system increases. So does the pituitary-adrenal, corticosteroid, and thyroid functions [54, 55]. Skeletal muscle tone increases, pulmonary permeability and respiratory volume of the lungs grow. Gaseous exchange intensifies with a subsequent growth in oxygen uptake by body tissues.

After moving from the continental climate to the seacoast, the basic metabolism may increase rapidly or gradually by 10–20% of the initial values during the first 2 months [56]. The activation of neurohumoral regulation and metabolic reactions in new and comfortable or almost comfortable conditions of the temperate marine climate is accompanied by a slight intensification of the cardiovascular activity: the contractile function of the heart increases and the coronary circulation improves, stroke and minute volume of the heart increase, the number of functioning skin capillaries increases, blood flow accelerates, and BP slightly decreases or increases in hypotensive states.

The processes described in the preceding text show that adaptation to the marine climate is complex in character and is determined by the diversity of physical characteristics of the climate of different geographic zones as well as by the chemical properties of the sea air. In temperate latitudes, the character of adaptive reactions constantly changes according to the seasons of the year.

The strongest stress reactions occur during a rapid change of climatic conditions [57]. In this period malfunctions of various physiological systems of the organism are often observed. During the initial period of adaptation, the operation of the nervous system may be inhibited and inhibitory processes may dominate. The activity of the sympathetic and parasympathetic nervous systems as well as the humoral pituitary-adrenal system increase. Pulse rate grows, systolic and diastolic blood pressure goes down, the volume of blood flow per minute increases, peripheral resistance reduces, and the value of the endurance coefficient of the cardiovascular system goes down. Respiration rate increases. Blood sugar falls, and the concentration of the total nitrogen in blood grows. There is a major deficiency of vitamins (B_1, B_2, B_6, PP, C), and there is a deterioration in work performance. It has conclusively been shown that on the fifth day these negative effects start to lessen, and on the 40–45th day after relocation, the first signs of adaptation and the rebalancing of major physiological functions are observed [58].

A relatively rapid change of climatic zones leads to constant adaptive stress for the body, resulting in the need for vitamins, a higher nutritional value of food, increased water and mineral intake, and a special dietary regime beneficial for a particular type of weather [59].

Chapter 5
Influence of Weather and Climatic Conditions on Health Adaptation to the Marine Climate of Russian Regions

5.1 Adaptation to the Marine Climate of the Black Sea Coast: Recuperation in Sanatoriums and Resorts

The marine climate of southern latitudes (the Black Sea coast of the Crimea and the Caucasus) is characterized by a large number of sunny days, high levels of solar radiation, especially of the ultraviolet spectrum, gentle winds, clean and fresh air, which is rich in ozone and sea salts, iodine, and sulfur compounds. All these factors reduce blood pressure and lead to an increase in protein and mineral metabolism. These climatic characteristics are beneficial for maintaining body heat balance, increasing the level of red blood cells, diuresis, secretion of endocrine glands, and metabolic activity. Data from several sources have identified a positive impact of the Black Sea climate on heart rate and lung ventilation. The constant movement of air masses produces an effect, which is similar to massage. All these factors shorten the period of adaptation and facilitate acclimatization [60]. This type of climate is well suited for the treatment of weaker patients since it is quite mild.

There are two climatic zones along the Black Sea coast of Crimea and the Caucasus: northern and southern. The northern zone stretches from Anapa to Tuapse, and the southern from Tuapse to the border with Abkhazia. The northern part of the coast, near Anapa, has a temperate continental climate. Summers are hot, but sea winds reduce day temperature, so one can hardly feel the heat.

The southern zone is an area of humid subtropics. The Caucasian Mountain range protects the coast from cold northern winds. At the same time, the sea warms the terrain. These factors create conditions for the formation of a humid subtropical climate. High mountains prevent dry continental winds from reaching the area [8]. The main characteristics of this climate are formed under the influence of several physiographic factors, the most important being solar radiation, atmospheric circulation, and relief. The Black Sea serves as a conductor of Mediterranean warmth: in

A. V. Tarasov, R. S. Rakhmanov, *Marine Climate of Russian Coastal Territories*,
https://doi.org/10.1007/978-3-031-30951-9_5

winter it has a warming effect on the North-Western Caucasus and in summer, the winds from the sea bring coolness and rain.

High mountain ranges act as a barrier, protecting the coast from cold continental northern winds. At the same time, they retain heat and moisture on the coast, thus increasing humidity and precipitation. In general, these conditions are typical of a subtropical climate on a narrow strip of the Black Sea coast.

5.1.1 Solar Radiation

Academic literature has revealed that the amount of solar energy received by a particular area is one of the factors determining the type of climate. In this regard, the southern coast of Crimea is located in a very favorable climate zone [61].

The duration of sunshine is 2200–2400 hours per year. The highest level of solar radiation is registered in summer, when the sun is close to the zenith, and solar radiation is about 15–18 kcal/cm^2. It drops in winter due to the shortening of the day and increasing cloudiness and is no more than 3–4 kcal/cm^2 per month. The radiation balance is positive during the whole year. This can be explained by a relatively high heat loss due to decreasing solar radiation, which can still reach 4 kcal/cm^2 per month even in winter. The maximum annual radiation balance is 50.0–52.5 kcal/cm^2 (in the northwest) and 52.5–55.0 kcal/cm^2 (in the southeast) [21].

The southern coast of Crimea has a high level of ultraviolet radiation coming not only from the sun, but also from the entire celestial vault. Even when the sun is at its highest point, the ultraviolet radiation of the cloudless sky exceeds the solar radiation by 10%. Ultraviolet rays stimulate the formation of enzymes and have bactericidal and anti-inflammatory properties [1].

5.1.2 Air Temperature

The average annual temperature of the Black Sea coast ranges from +11.8 (Anapa) to +14.1 °C (Sochi). The average temperature of the coldest month of the year, January, is above zero and ranges from +1.2 °C (Anapa) to +5.8 °C (Sochi). The absolute minimum temperature was registered in Novorossiysk (−24 °C) and Sochi (−15 °C). Due to the cooling effect of the sea, the average July temperature is +23 to 25 °C, and the maximum temperature is usually registered in August. It fluctuates between +23 to 23.2 °C in Anapa and +22.8 to 23.3 °C in Sochi. Absolute maximum temperatures were registered in Tuapse (+41 °C) and Sochi (+39 °C). Winter temperatures are rarely below 0 °C and a decrease in air temperatures to negative values is usually observed only for a few days. The frost-free period lasts 220–260 days.

5.1.3 Cloudiness

Observations of cloud cover show that fogs and laminar clouds are most often formed at night and in the morning hours, when the temperature is low and the humidity is high. Cumulonimbus and other types of clouds are observed in the warm season and develop most frequently in the afternoon hours, when the convection is at its strongest. During the warm season, there are two maxima in the diurnal progress of cloudiness: in the morning and in the afternoon, the latter being more intense. In the cold season, maximum cloudiness is observed in the afternoon. In winter, the frequency of overcast conditions is 65–70%. In summer months and in September, the probability of clear and partly clear sky is about 80%, and in August 85%.

The abundance of heat and light, which the South Coast of the Crimea is famous for, is explained not only by its southern location, but also by the exceptional purity and transparency of the atmosphere, easily permeable to all types of solar rays. The cloud coverage is not extensive, not more than 50%, that is, clouds cover not more than half of the sky. In summer, the cloud coverage decreases to 30%. In Yalta, there are 276 sunny days a year. The number of clear days is 128, and there are only 52–56 days when the sky is overcast.

However, the sun can do harm to the body. Excessive exposure to the sun has a negative impact on the cardiovascular system, may aggravate chronic diseases and cause bleeding, etc.

5.1.4 Air Humidity

High humidity decreases water excretion by the skin and lungs, increases urine excretion, and has adverse effects on patients with kidney diseases. A rapid increase in humidity can cause bleeding. On rainy days patients complain of increased shortness of breath, heart palpitations, headaches, "foggy" brain, fatigue, apathy, pain in different parts of the chest, and increased amount of sputum.

Precipitation (rain, snow, etc.) contributes to air purification and is not considered harmful in climatotherapy, unless rainy or snowy periods are frequent or lasting long enough to preclude staying outdoors. Frequent rains have a negative effect on the psyche and mood, especially in those with an unstable nervous system. Dry and warm air reduces urination, so it is favorable for patients with kidney disease. But excessively dry air is harmful to the body: it irritates airways, causes general anxiety, headaches, insomnia, and intensifies coughing.

The relative humidity in the warm period of the year is 69–71% (Anapa), 65–68% (Abrau-Durso), 63–64% (Novorossiysk), 65–68% (Gelendjik), 74–76% (Dzhubga), 70–71% (Tuapse), and 76–77% (Sochi). The highest values of relative humidity are observed in the cold period (December–January): 80% (Anapa), 80% (Abrau-Durso), 77% (Novorossiysk), 72% (Geolendzhik, maximum in May – 76%), 79% (Dzhubga, maximum in May – 80%), 71–72% (Tuapse, maximum in May–June – 76%), and 70–72% (Sochi, maximum in May–June – 78%).

5.1.5 Precipitation

The average annual precipitation across coastal resorts ranges from 452 in Anapa to 1543 in Sochi. The Black Sea has a marked effect on the annual precipitation cycle, with a maximum in the winter and a minimum in the summer. The total annual precipitation decreases northward: it reduces from 1.000–1.500 mm in the south of the Black Sea coastal zone to 600–800 mm in the north. Transitional areas, where both summer and winter precipitation maxima occur, are along the coast of the Sea of Azov and in the foothills of the Caucasus.

Dry summer air takes the edge off the heat. The moderate humid climate of the southern coast of Crimea, which has little precipitation (540 mm on average with 83 rainy days) and a narrow daily and seasonal precipitation range, is good for health.

5.1.6 Seasonal Climate Changes

The climate of an area is best described through seasonal changes. On the Black Sea coast, temperatures rarely drop below 0 °C, and there is no winter as understood by agrometeorology. Rainfall is heavy from December through February. Continuous downpours are usual during half of the period, while the snow cover is present only a few days.

Frost periods are light and short, and they do not occur each year. Cold, snowy winters happen once in 7–11 years. And when they do, temperatures do not fall below 5–10 °C. In December–January, when severe frosts set in the north of Russia, the temperature in Yalta is usually about +4 to 6 °C.

Novorossiysk has 20 days with average daily temperatures below zero; Sochi, only four. Yet some winters see cold spells down to −14 °C. In Anapa and Novorossiysk, the absolute minima are at −24 to 26 °C.

Springs on the southern coast are early but long. Sea waters, having grown cold during the winter, cool the air. Temperatures rise in March when cold winds blow. By the end of April, the temperatures of the air and sea waters even up. The sun shines brighter and gives off more heat; trees are in full blossom. In May, the temperature of the air is higher than that of the sea.

It rapidly warms up in spring. Fifteen days into the season, in March, the air temperature rises above +5 °C to reach +10 °C and get even higher on 10–20 April (or in the first days of April on the coast). The temperature does not drop below zero any more (in Sochi, this happens at the beginning of March).

The summer begins in the first half of May. The season is moderately hot. The temperature in July is +21 to 22 °C with maxima of +35 to 38 °C or even +40 to 43 °C in flatlands. Proximity to the sea lowers air temperatures on land in warm seasons and prevents the cooling of the air at night and in cold ones. Thus daily, monthly, and annual temperature variations are evened out on the southern coast of Crimea. The summer in the area is dry and warm, but not hot.

Heavy rains account for most summer precipitation. The Black Sea coast has 700–800 mm of rainfall from April to October; the northern regions, only 200–300 mm. The area has short periods of sukhovey winds, which do not exceed 10–15 days in the summer. During these days, precipitation is low.

5.1.7 Atmospheric Pressure

The long-term average shows that the pressure is slightly higher in the winter than in the summer. The average annual pressure is 1020.2 and 1011.0 mbar in Anapa and 1019.8 and 1010.7 mbar in Sochi. Atmospheric pressure maxima are observed in the winter, in January; minima, in the summer (July). The average annual air pressure is 1015.9 mbar in Sochi and 1016.2 mbar in Anapa. The standard deviation is within 0.1–0.3 mbar. The daily pressure variation in Sochi is 38 mbar. Daily variations in the summer are more subtle than in the winter. The southern coast of Crimea has slightly higher atmospheric pressure, yet its variation is slight, and thus it does not harm the human body.

5.1.8 Atmospheric Circulation and Wind Direction

The mountain ridges of the Caucasus affect the wind regime and create a local mountain-valley circulation determined by pressure differences, which result from changes in temperature during the day. In the afternoon, the air warms and rises toward the mountain peaks while cooled air descends the slopes at night. The direction of the wind usually changes between 8–10 am and 8–10 pm. In the summer, mountain-valley winds bring air from lower-lying valleys to slopes higher in the mountains where water vapor condenses, turns into cumulus clouds and precipitation, and prevents a rise in the temperature. At night, the winds blow in the opposite direction and cool the air less thoroughly than they do in the valley. Mountain-valley winds are most characteristic of the northern coast of the Black Sea. On open capes and spits protruding into the sea and on uplands, these winds blow 35–60 days a year.

Prevailing winds come from the north-east. The wind is the strongest at the Markotkh Pass. The average annual wind speed in Sochi is 2.8 m/s, while it is 3.8–8.8 m/s at the pass. When anticyclones settle over the European part of Russia, the Novorossiysk bora – a cold gusting north-east wind – prevails at the western edge of the Caucasus, on its Anapa-Novorossiysk-Tupase stretch in the colder seasons. In the warmer seasons, this wind blows much more rarely. Air moves from the anticyclone toward the quasi-stationary Black Sea depression, which is a low-pressure area. The wind, having crossed the low ridges, sweeps the Tsemes Bay and the adjacent coast. In Novorossiysk, the bora prevails during 40 days; gusts reach up to 50 m/s. High north-east winds start to blow across the sea when anticyclones settle over the Krasnodar region, and the pressure is low. They may produce the force of a hurricane. Prevailing winds are from the south-east. The north-east bora is frequent at the end of the winter and in the early spring, from February to March. In February, storms are the strongest; wind speeds reach up to 18 m/s. In some years, boras blow during 10–15 days.

Local circulation comprises breezes and föhns. Breezes blow from the sea toward the land in the daytime and from the land toward the sea at night. They remain stable when anticyclones prevail. These winds are usual on the Black Sea coast in the summer and spring. The direction from which the sea breeze originates changes from south-east to south-west. Land breezes coming from the mountains are föhns blowing from north-east. They supersede sea breezes at 7–10 am in the summer. The reverse change occurs from 6 to 9 pm. The vertical thickness of the sea breeze is 800–1000 m; of the land breeze, 250 m. The sea breeze reduces the heat. Breezes are most common in the summer in anticyclone weather, which is never affected by fronts passing through or air masses changing. Weaker breezes blow on the coast of the Azov Sea and near large reservoirs.

Local mountain-valley winds and breezes have a beneficial effect on the human body. Breezes bring down the heat in the daytime and make nights warmer. The sea breeze carries toward the land clean cold air rich in sea salts and ozone. The land breeze and the mountain-valley breeze bring from mountain terraces aromatic ozone-rich air filled with resinous substances and essential oils. Therefore, the southern coast of Crimea has a very favorable wind regime.

Föhns are descending winds produced by temperature and air humidity contrasts between the leeward and windward mountain ridge slopes. These dry winds blowing from mountains bring a sharp rise in the temperature and a complete or partial disruption of daily temperature and air humidity patterns. Föhns are most frequent in the spring. In the autumn and particularly the winter, they dramatically affect the climate by reducing air humidity and increasing temperatures in the winter months. The speed of föhns is 15–20 m/s, sometimes reaching up to 25–30 m/s. The duration of winds ranges from several hours to 4–5 days. The average annual wind speed varies enormously. It is 6.03 m/s in Anapa, 3.52 in Abrau-Durso, 4.64 in Novorossiysk, 3.34 in Gelendzhik, 5.1 in Dzhugba, 4.42 in Tuapse, 2.8 in Sochi, and 4.5 in Adler. On the coasts of the Black and Azov Seas, marked differences between thermal and pressure patterns result in higher wind speeds in Dzhugba (5.1 m/s), Yeysk (5.9), and Dolzhanskaya (6.6 m/s).

5.1.9 Ionization and the Chemical Composition of Air

The high air ionization rate adds to the therapeutic benefits of the benign climate of the southern coast. The therapeutic effects of marine and mountain climates have been attributed to high air ionization.

As biologically active particles, negative ions have a beneficial effect on pulmonary ventilation. Moreover, they stimulate oxygen uptake and carbon dioxide production, increase the concentration of red blood cells and hemoglobin in the blood, slow down the heart rate, and accelerate metabolism. Negative air ionization therapy has been shown to improve well-being, enhance mental and physical stamina, normalize sleep and appetite, stimulate growth and development, and activate body defenses.

Clean marine air, which is almost free of dust and microörganisms, has a positive effect on sufferers of chronic upper airway and pulmonary diseases. Marine air has an optimal oxygen-to-ozone ratio, improves metabolism, and normalizes breathing patterns [60].

Some chemical elements contained in the air benefit health. These are phytoncides, terpenes, and sea salts. Plants can also release substances destroying disease-causing pathogens, including phytoncides and terpenes.

Alongside the antimicrobial activity, phytoncides are known for their beneficial effect on mental health, improving the breathing function, normalizing heart rate and blood pressure, activating tissue respiration and energy exchange, and reducing inflammation. The air in the region is heavy with the scents of herbs and flowers. The smell of roses has been reported to reduce fatigue; laurel, carnation, and iris, to tone up the nervous system; lavender, mint, sage, and anise, to enhance cerebral blood flow [60].

Sea salts turn the coast into somewhat of a gigantic natural inhalation room. They impart therapeutic properties to the air and make staying on the Black Sea coast even more beneficial to health.

Comprehensive physiological studies have shown that the marine climate of the Black Sea coast of the Caucasus has a positive impact on the functional reserves of the cardio-respiratory system and increases the efficiency of oxygen uptake during exercise [62].

The therapeutic properties of the hot marine climate of Anapa have been exploited in improving children's health. The local environment has been shown to produce an adjuvant therapeutic effect and increase non-specific immune resistance in children suffering from mucopurulent chronic bronchitis. After thalassotherapy, normalization of T helper and T suppressor cells occurred in patients of the treatment group. At the same time, immune status indicators were optimized, including IgA, IgG, IgM (identified in standard alveolar lavage) and IgE (determined in serum). The activation of non-specific immune resistance in children suffering from chronic diseases of lower respiratory diseases manifested itself in the normalization of helper-suppressor cell ratio [25].

The rationale behind this methodology, namely, the frequency, repetition, duration, seasonality, and compatibility of different procedures used in treating children at Anapa resorts, has been described in the literature [63]. In some months, high humidity and storms beset the southern latitudes and create conditions that are not beneficial to sufferers of some diseases. It has been demonstrated that among visitors from Russian northern and southern regions, 32% and 19%, respectively, experience maladaptation upon arrival at the Black Sea coast. Although northern visitors undergo a more dramatic and effective restoration of the functional reserves of the cardio-respiratory and metabolic systems than their southern counterparts do, the rehabilitation of the former has some specific features, and the therapeutic effect comes at an "acclimatization price" [64].

Over half of the children feel unwell in the first days in the south. In particular, discomfort, undesirable behavior, fatigue, headaches, insomnia, and a loss of appetite have been observed. Youngsters have been diagnosed with dysautonomia, fluctuating blood pressure, ventilation and temperature regulation problems, and adynamia. Children from environments similar to the Black Sea coast adapt more easily than their peers from the northern regions and other territories with a different climate. Children from the north and industrial cities show signs of ultraviolet light deficiency and are very susceptible to solar radiation.

Residents of central Russia feel overheated when the heat index, which combines air temperature, relative humidity, and wind speed, is above 23 °C. Summer daytime temperatures are usually about 28–30 °C on the Black Sea coast. Therefore, visitors at southern resorts are in danger of overheating, which has an adverse effect on the cardiovascular and nervous systems.

Spending too much time in the sun leads to sunburns and overheating with various clinical manifestations, including heatstroke. The clinical symptoms are overexcitability, fatigue, apathy, lethargy, headaches, and sleep disorders. Sometimes they are accompanied by ear, nose, and throat inflammation. Children with chronic diseases may experience flare-ups. In the summer, the number of calls for an ambulance for children rises, and hospitals in resort towns are rammed with children, many of whom are tourists suffering from conditions associated with excessive solar radiation.

Ultraviolet excess is typical of areas south of the latitude 42° north – the Mediterranean, the Red Sea, and Transcaucasia. On the Russian Caucasian coast and at the resorts of the North Caucasus, midday hours are the most dangerous. At that time, ultraviolet excess, accompanied by increased biological solar activity, may lead to cancers, solar allergy, and the exacerbation of cardiovascular diseases, diabetes, and liver and renal disorders.

5.2 The Effect of the Black Sea Climate on the Human Body During Job Adaptation

Our study was carried out in Anapa, which lies on the north-east coast of the Black Sea at 44° 53′N, 37° 19′E, and has a submontane-steppe Mediterranean climate.

A group of young men aged 18–19 was selected for the study. Subjects were divided into two categories based on anthropometry: those with normal nutritional status (BMI above 19.4 kg/m^2) and poor nutritional status (BMI between 19.4 and 18.5 kg/m^2). During the study, all subjects received the same nutrition. The average seven-day daily calorie intake was 3675 kcal. The diet met the energy needs of people performing heavy physical work. The young men spent over a third of the daily time budget (34% or 490 min) working in the open air in the daytime.

In warm seasons, the wet bulb globe temperature (WBGT) was 29.8 ± 0.34 °C. Some days, it reached values hazardous to people unaccustomed to a southern climate (31.1 °C) [65]. The intensity of open-air work did not change as the temperature rose. The WBGT-index was used to evaluate the effects of temperature on the human body. The WBGT combines convective and radiative temperatures, humidity, and wind speed, measured with a black globe thermometer. At the WBGT above +26.7 °C, avoiding extensive physical exercise in the daytime is recommended.

In the cold seasons, the total cooling effect of climate factors was determined using the wind chill index (WCI). The WCI was 960–1089 kcal/m^2/h (the comfort threshold is at 770 kcal/m^2/h or 3200 kJ/(m^2/h)). This index evaluates the microclimate of a cold environment [66].

In 3 months, the BMI of subjects was decreasing, with the most dramatic reduction taking place on day 41 when the BMI went down by 1.1 ± 0.2 kg in group 1 and by 0.7 ± 0.1 kg in group 2. In 70% of subjects in group 1, body weight dropped by 1.5–3.5 kg; in 90% of group 2, by 2.2–4.9 kg. By the end of the study, body weight reverted to initial values in group 1, while it remained at the level of stage 3 in subjects with poor nutritional status (Fig. 5.1).

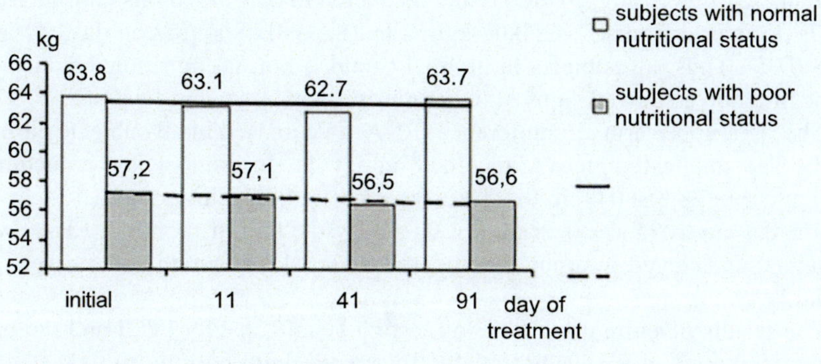

Fig. 5.1 The BMI of treatment subjects, absolute values

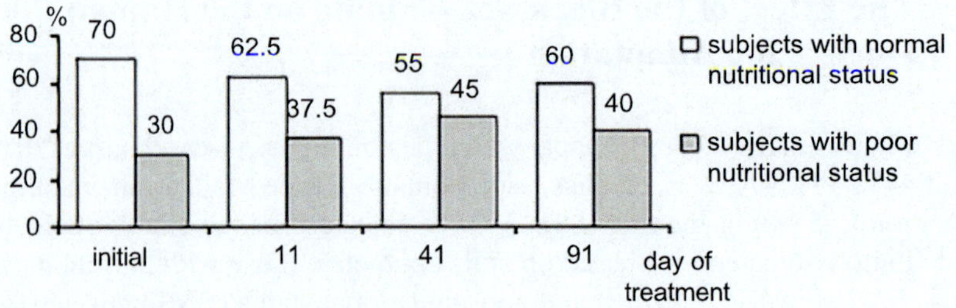

Fig. 5.2 Subjects experiencing weight loss, %

Table 5.1 The structure of body weight change in subjects with normal (1) and poor (2) nutritional status during treatment, %

	Increase			Decrease			No change		
	Day of the study								
Treatment group	11	41	91	11	41	91	11	41	91
1	10.6	7.1	10.2	78.8	89.3	79.0	10.6	3.6	10.8
2	33.3	25.0	25.0	50.0	75.0	69.0	16.7	–	6.0

The dramatic body mass reduction observed on day 41 was accompanied by a 1.3-fold decrease in the number of subjects with normal nutritional status and a 1.5-fold increase in the number of those with poor nutritional status (Fig. 5.2).

An evaluation of body weight changes in the experimental groups showed that only 10.6% of males in group 1 (Table 5.1) gained weight by day 11 of the study, while the weight of more than three-fourths of the subjects decreased. By day 41, the weight of every ninth person dropped; 3 months later (on day 91), the figures matched those obtained on day 11 of the study.

A third of underweight subjects gained weight after 10 days of a sufficiently high-calorie diet. Yet half of them lost weight over the period. After 40 days, weight gain was observed in 75% of subjects; weight loss in 25%. This ratio did not change significantly during the 3 months of the study.

BMIs decreased in group 1 from 21.0 ± 0.3 kg/m2 to 20.6 ± 0.3 kg/m2 by the third phase of the study ($p > 0.05$) and recovered to baseline by its end. In group 2, BMIs decreased from 18.7 ± 0.07 kg/m2 to 18.3 ± 0.08 kg/m2 on days 41 and 91 ($p > 0.05$). Thus, all subjects in group 1 retained normal nutritional status, while some in group 2 showed signs of undernourishment.

The right upper arm circumference (RUAC) decreased in all subjects during the study. The smallest value was recorded on day 41. In group 1, the reduction was 1.7 cm (from 28.8 ± 0.4 cm to 27.1 ± 0.4 cm, $p < 0.05$); in group 2, 1.3 cm (from 26.4 ± 0.4 cm to 25.1 ± 0.3 cm, $p < 0.05$). By the end of month 3, RUAC values remained unchanged in group 2 compared to day 41; in group 1, they were at the initial level.

The results of pulmonary function testing (vital capacity [VC] and the breath-hold test) did not change substantially. In subjects with poor nutritional status, the

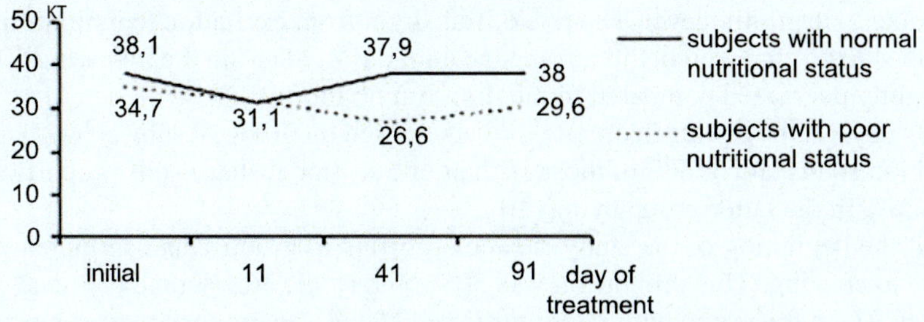

Fig. 5.3 Changes in the dominant hand strength in the experimental groups, kg

Table 5.2 Percentage of subjects with a reduced rate of transition from excitation to inhibition, %

	Day of the study			
Treatment group	Initial	11	41	91
Subjects with normal nutritional status	63.3	63.3	63.3	52.0
Subjects with poor nutritional status	67.9	71.4	71.4	68.1

VC decreased by day 11 in 33.3% of cases (1.4 times as frequently as in group 1). In this group, a 35.7% reduction in breath-holding time occurred. These changes persisted until the end of the study.

In the group with poor nutritional status, the dominant hand strength declined from 37.1 ± 2.5 kg to 36.8 ± 2.8 kg ($p > 0.05$). In group 1, this parameter did not change (38.6 ± 1.7 kg).

On day 11, muscle strength decreased in 25% of subjects with poor nutritional status (as in group 1); by day 41, this proportion increased 1.7-fold. In young males with normal nutritional status, the rise was 1.3-fold. Moreover, the period of decline in muscle strength was shorter in the latter group: on day 41, there was no significant difference from baseline. In group 2, the decrease in muscle strength was substantial throughout the study (Fig. 5.3).

Throughout the initial period of adaptation, the performance of the tapping test did not vary significantly in subjects with normal nutritional status, while subjects in the other group scored lower than they did at the beginning of the study (196.0 ± 4.0 points compared to 210.1 ± 7, ($p < 0.05$)). Individual performance worsened in the group with poor nutritional status 1.4 times as much as it did with normal nutritional status. By day 91, the proportion of subjects with lower test performance increased to 67.4% (2.1-fold) in group 2, whereas it remained unchanged in group 1 (32.1%).

No significant changes in the rate of visual information processing occurred. Yet, a slight decrease was observed in group 2, from 0.93 ± 0.08 bps to 0.89 ± 0.05 bps on day 91, $p > 0.05$. Among subjects with normal nutritional status, the percentage of individuals with a reduced rate of transition from excitation to inhibition was 1.2 times that in the group with normal nutritional status (Table 5.2).

As the table shows, the percentage of subjects with signs of nervous system fatigue increased by day 11 and remained at this level until day 41; on day 91, it returned to the initial level. The rate of transition from excitation to inhibition in the group with normal weight did not change until day 41. During the following 50 days, it slightly decreased compared to the first examination.

By day 11, the resting heart rate (HR) increased in 39.3% of subjects with normal nutrition status and 59.9% of those with poor nutritional status (this proportion rose to 76.6% in the latter group by day 91).

At the beginning of the study, 50% of subjects in group 1 had an increased HR while exercising. This proportion was 50% on day 11, 36.1% on day 3, and 32.1% on day 91. In the group with poor nutritional status, the proportion of subjects with increased HR was 65% on day 11, remained unchanged until day 41, and decreased to 61.3% by the end of the study (Fig. 5.4).

Heart rate recovery (HRR) during the standard time after exercise was slower in subjects with poor nutritional status (Figs. 5.5, 5.6).

At the first stage of the study, symptoms of cardiovascular tension were observed in 41.7% of subjects in group 1 and 42.4% in group 2 when performing graduated exercise (Table 5.3).

The proportion of subjects with an abnormal cardiovascular response was the largest on day 11 in group 2 and day 41 in group 1. No improvement occurred until the end of the study.

Urinary excretion of ascorbic acid measured in mg per hour at the beginning of the study showed that vitamin C saturation in group 1 was 1.2 times that in group 2.

In a considerable proportion of subjects, the values of ascorbic acid excretion were either below the normal range (<0.4 mg/h) or at its lower end (0.4–0.5 mg/h): 50 and 28.6% in group 1; 58.3 and 25% in group 2. The level of normal vitamin C saturation decreased in both groups (Figs. 5.7, 5.8).

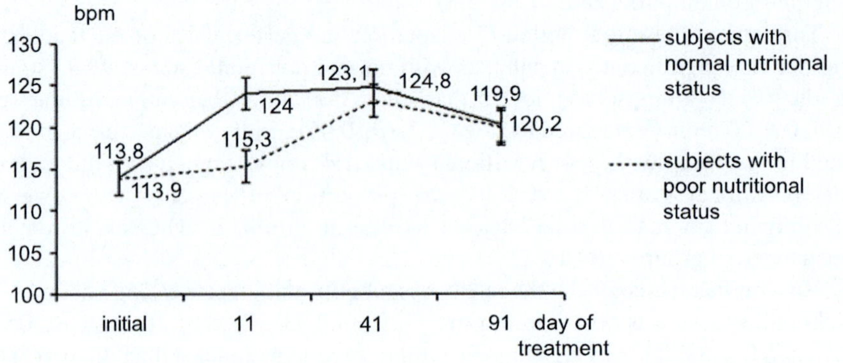

Fig. 5.4 HR while exercising in subjects with different nutritional status, bpm

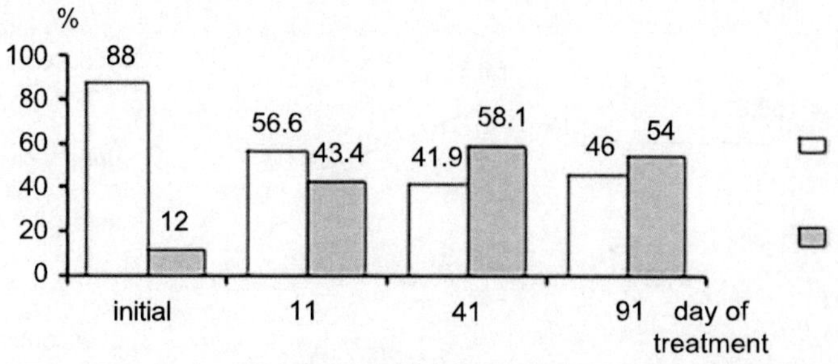

Fig. 5.6 Percentage of subjects with different HRR after graduated exercise (poor nutritional status), %

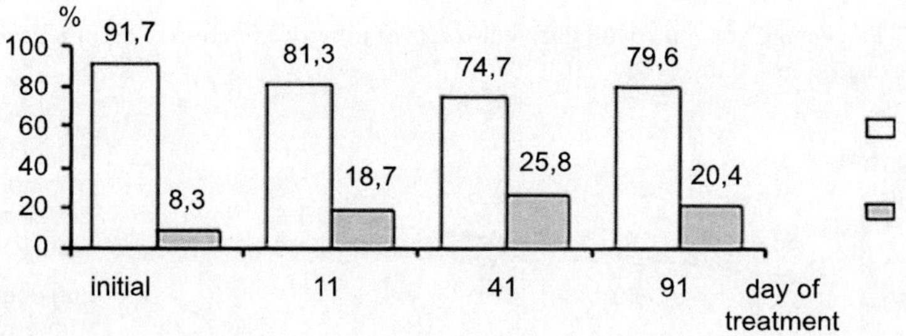

Fig. 5.5 Percentage of subjects with different HRR after graduated exercise (normal nutritional status), %

Table 5.3 Percentage of subjects with abnormal response to graduated exercise, %

	Day of treatment			
Treatment group	Initial	11	41	91
Subjects with normal nutritional status	41.7	60.7	66.7	65.8
Subjects with poor nutritional status	42.4	75.0	71.4	74.2

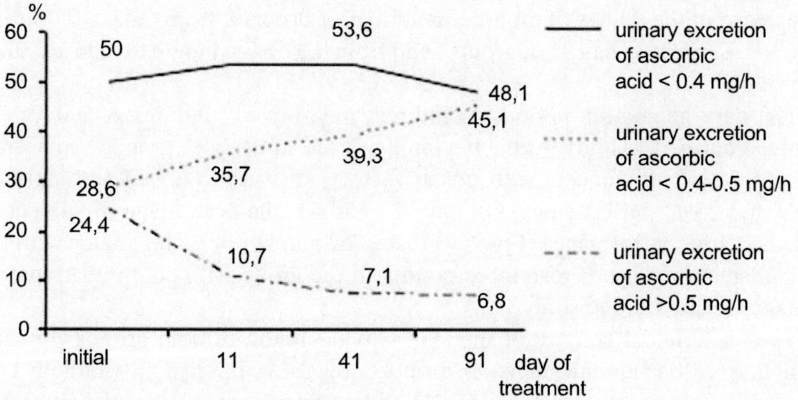

Fig. 5.7 Percentage of people with different levels of urinary excretion of ascorbic acid (subjects with normal nutritional status), %

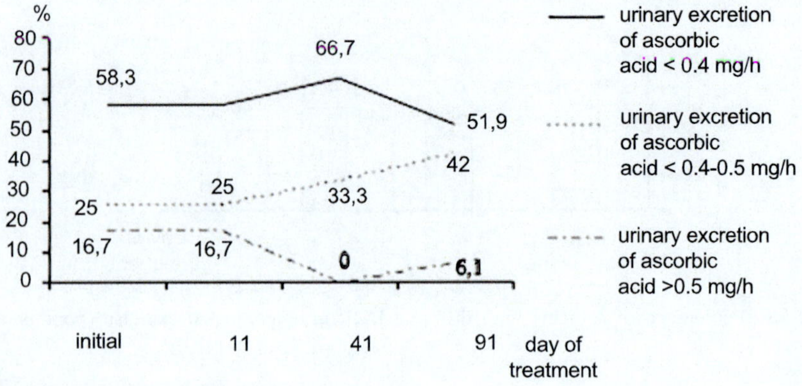

Fig. 5.8 Percentage of people with different levels of urinary excretion of ascorbic acid (subjects with poor nutritional status), %

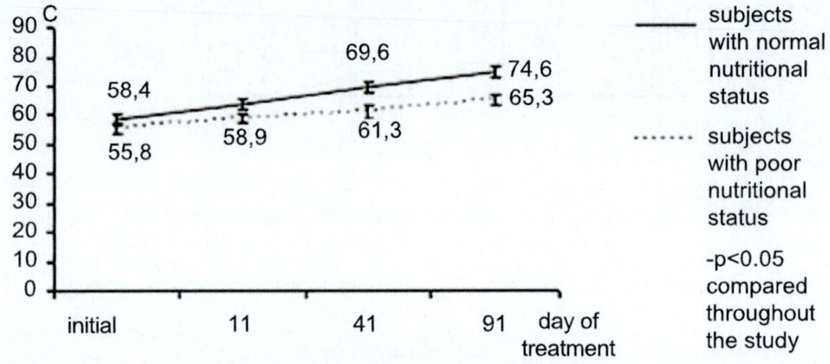

Fig. 5.9 Changes in dark adaptation periods in experimental groups, s

Subjects with normal vitamin C levels at the beginning of the study experienced a decrease in the values during the treatment: it dropped from 0.43 ± 0.03 mg/h to 0.36 ± 0.02 mg/h on day 91 in group 1 and from 0.37 ± 0.04 mg/h to 0.33 ± 0.02 mg/h in group 2.

The dark adaptation period, an indirect measure of vitamins A and B_2 levels, changed during the study. At the beginning of the study, both groups had a substantial proportion of subjects with unsatisfactory (57.1 and 66.7%) and satisfactory (39.3 and 25%) performance. On days 11 and 41, the percentage of subjects with unsatisfactory performance (Fig. 5.9) rose. The number of young males with longer dark adaptation periods rose more rapidly in the group with poor nutritional status at stages 2 and 3 of the study.

The bactericidal activity of the skin was declining in both groups. In subjects with normal nutritional status, it dropped by 3.3% upon a five-minute contact between skin and cell culture on day 41 compared to the initial levels ($p > 0.05$); by

Table 5.4 Subjects with mild (D_1) and more severe (D_2) dysbiosis in subjects with normal and poor nutritional status, %

	Day of treatment			
Treatment group	Initial	11	41	91
With normal nutritional status:				
Normal	42.9	23.8	28.6	25.1
D_1	23.8	42.9	42.9	44.5
D_2	33.3	33.3	28.6	30.4
With poor nutritional status:				
Normal	33.3	–	–	–
D_1	33.1	33.3	11.1	48.7
D_2	33.6	66.7	88.9	51.3

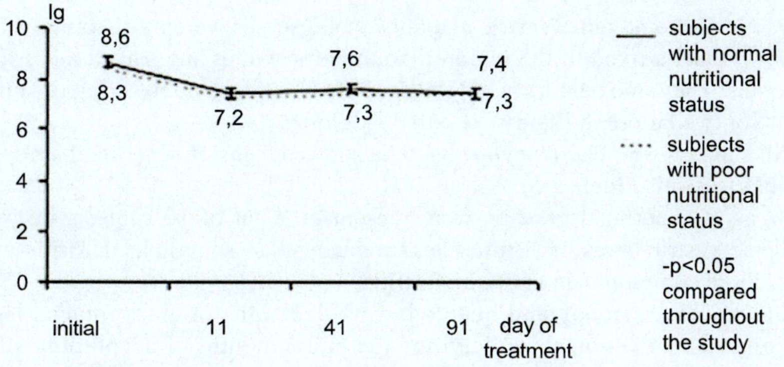

Fig. 5.10 Presence of *Bifidobacteria* in feces, lg/g

10.7% ($p < 0.05$) upon a 10-min contact. By day 91, the bactericidal activity of the skin recovered.

In the group with poor nutritional status, the bactericidal activity of skin upon a 5-, 10-, and 15-min contact with a cell culture tended to drop as early as day 11. From day 41 to day 91, a significant decline in skin bactericidal activity of skin was observed.

At the beginning of the study, the composition of large intestine microbiota in subjects with normal nutritional status had signs of dysbiosis. Later, the proportion of individuals with mild dysbiosis increased, while the percentage of subjects with normal microbiota declined (Table 5.4).

The proportion of individuals with more severe dysbiosis rose in the group with poor nutritional status to reach maximum values on day 41.

The level of *Bifidobacteria* decreased by day 11 in both groups; it did not change until the end of the study. The proportion of subjects with a low level of bacteria increased twofold during stage 2 (by 57.1%) in group 1 and threefold (by 66.7%) in group 2 (Fig. 5.10).

Table 5.5 The histology of the gastric antral mucus in the presence of *H. pylori*

Type of gastritis	Absolute values	%
Superficial	115	89.8
Atrophic	10	7.8
Granular	3	2.4
Total	128	100

The incidence of gastritis, duodenitis, and other diseases of the stomach and duodenum was increasing. Most subjects suffering from these conditions were adapting to the local weather and climate (0–6 months, 41.8%). Peptic ulcer was 1.5 times more frequent in newcomers than in individuals who had not resettled in the past 18–24 months.

Peptic ulcer and other gastric and duodenal diseases were most common in subjects who had arrived in the region in autumn or winter and stayed there for 2 or 3 months. The same held for young males who had relocated in spring or summer a month or two before the study.

All subjects who had *H. pylori* bacteria showed signs of superficial, atrophic, or granular gastritis (Table 5.5).

Signs of superficial gastritis were present in 9 out of 10 subjects in group 1. Atrophic gastritis was 11.5 times less frequent than superficial gastritis and 3.3 times more common than granular gastritis.

Gastroduodenoscopy and histology revealed signs of antral mucosa inflammation in 92% of subjects undergoing the first 6 months of adaptation; stomach and duodenum inflammation in 40%; over 50 *H. pylori* per HPF in smears in 92% [62].

Weather and climate were risk factors for community-acquired pneumonia (CAP) in young men at early stages of adaptation. In the spring and summer, 4.5% of subjects contracted CAP during their first 3 days in the region; 9% on days 4–10; 10.5% on days 10–20; 6.5% after 20 days. As to the monthly distribution, 30.4% occurred in the first month; 46.9%, the second; 18.2%, the third; 4.8%, the fourth [53].

In the autumn and winter, 2.6% of CAP cases were diagnosed in the first week; 5.6%, the second; 12.3%, the third. The monthly distribution was as follows: 21% occurred in the first month; 26.2%, the second; 29.2%, the third; 23.6%, the fourth.

Analysis of integrated climate indices revealed correlations between CAP incidence and the WBGT in the summer ($r = 0.67$, $p < 0.01$) and between CAP incidence and the WCI ($r = 0$) in the winter ($r = 0.67$, $p < 0.05$). Climatic factors had a very similar influence on CAP incidence among subjects in both seasons. The weight of climate factors was 15.1% in the winter and 15.2% in the summer, based on the R squared of a linear regression model.

5.3 The Effect of the Caspian Sea Climate on the Human Body During Job Adaptation

The climate in the region is arid. Winters are mild. In the summer, the precipitation-evaporation balance varies from 320 mm in the north to 380 mm in Makhachkala. Summers are hot. The average temperature in the hottest month is +28 to 29 °C. The absolute maximum for Makhachkala is +37 °C. The relative humidity is 63%. The amount of precipitation in the summer is only 60–70 mm.

In the autumn, the temperature gradually decreases to drop below zero in the middle of November. Precipitation in the autumn is heavier than any other time of the year. The precipitation-evaporation balance is 16 mm. In November, this ratio is higher. The ice-free Middle Caspian Sea affects the climate in the winter.

The average January temperature is about 0 °C–0.4 °C. The temperature drops below −9 °C every year. The absolute minimum for Makhachkala is −23.9 °C. In some years, temperatures rise as much as +10 °C in January. The influence of the sea is the strongest in the southern parts of the region. January temperatures range between −1 °C and +0.5 °C; the average low is −13 °C. Precipitation in cloudy winters is low; the unstable snow cover does not exceed 7 cm.

The influence of the sea is also significant in the spring when the water and air temperatures differ: the air above the land warms up quickly, while the air above the sea remains cold. This temperature variance results in a steady easterly wind.

The highest humidity and lowest air temperatures occur in the winter when humidity is 7–16.6% higher than in the summer, spring, and autumn. Winter temperatures are 20.57 °C lower than those in the summer; 7.07 °C, than in the autumn; 10.97 °C, than in the spring (Table 5.6).

The maritime climate of the Caspian Sea had the most adverse effect on health in the summer when maximum daytime temperatures reached +35 to 37 °C.

Like in the Black Sea area, adaptation to the weather and climate of the Caspian Sea negatively affected health. Bakhmudov [96] links acclimatization to unwelcome changes in the morphofunctional state (body weight, hand strength, and the condition of cardiovascular and respiratory systems). The thermal signs of immune deficiency diseases were detected in 86.8–96.7% of subjects [69].

A hot and humid maritime climate caused the proportion of persons with a decreased bactericidal activity of skin to rise by 15.2% [70].

A comparison of microorganism content in the feces of subjects arriving in the spring and autumn showed that the concentration of *Bifidobacteria* in the former

Table 5.6 Seasonal temperatures and humidity

Season	Air temperature, °C	Relative humidity, %	Wind speed, m/s	Atmospheric pressure, mm Hfg
Winter	3.53 ± 0.46	87.3 ± 0.51	3.1 ± 0.72	767.95 ± 0.43
Spring	10.6 ± 1.15	80.3 ± 0.62	2.8 ± 0.51	764.39 ± 0.55
Summer	24.1 ± 0.57	70.7 ± 0.75	2.4 ± 0.51	760.22 ± 0.29
Autumn	14.5 ± 1.13	80.3 ± 0.81	3.3 ± 0.68	767.1 ± 0.72

Table 5.7 Skin and subcutaneous tissue infections among young men, depending on the time spent in the Caspian Sea area, ‰

Time spent in the area, years	Morbidity	Excess factor of the first period
0–0.5	544.6 ± 4.76	–
0.5–1.0	276.2 ± 5.2	1.97
1.0–1.5	275.3 ± 3.57	1.98
1.5–2.0	269.5 ± 3.47	2.02

group had decreased in the initial adaptation period. In particular, the content of enterococci was significantly higher than normal levels. According to individual gut flora profiles, 50.0 ± 5.9% of subjects had D_1 dysbiosis. It was also diagnosed in 38.3 ± 6.4% of those who had arrived in the region in the autumn.

The weather and climate conditions affected morbidity among subjects. The most common categories of disease were skin and subcutaneous tissue conditions and respiratory disorders. The incidence of skin and subcutaneous tissue diseases was statistically higher than that of respiratory conditions by 11.3%. It was also 3.1–3.3 times that of infectious and parasitic diseases and digestive diseases [71].

When the group was set up in the spring, maximum morbidity was registered in months two and three. The first 3 months accounted for 88.8% of all morbidity events in the first 6 months.

The highest morbidity rate was observed in the summer and autumn; the lowest, in the spring. Long-term annual averages were very similar.

The incidence of skin and subcutaneous tissue infections among subjects who had arrived in the region within the past 6 months was 1.97–2.02 times that among people who had lived there longer (Table 5.7).

Furuncles, abscesses, phlegmons, and pyoderma are the most common conditions in the category of skin and subcutaneous tissue diseases.

Long-term data show a trend toward an increase in the number of skin and subcutaneous tissue infections. The annual increment was 25.4 per mille, and the increase rate was 6.9%. Growing morbidity is explained by a rise in the cases of pyoderma, furunculosis, abscesses, mycoses, and complicated pyoderma.

The most prevalent condition was pyoderma, which was diagnosed 1.4 times as frequently as furunculosis and abscesses and 10.5 times more often than phlegmon. Pyoderma, furunculosis, and abscesses comprised 90.2% of all cases in the category.

5.4 Adaptation to the Maritime Climate of Northern Latitudes

The maritime climate of northern latitudes is associated with few sunny days, frequent winds, clean and fresh air, and heavy precipitation.

The Arctic, where the coast of the Murmansk region lies, has a wide temperature range. A principal factor is the wind regime with frequent, strong, and gusty winds of different directions. The average annual wind speed is 5–8 m/s. The annual wind speed pattern suggests that stronger winds blow in the autumn and winter. On the coast of the Kola Peninsula, they reach a speed of 30–35 m/s. The constant inflow of moisture from surrounding seas results in a constantly high relative air humidity, which is a negative factor.

Climatically, all Arctic territories belong to the frigid zone. They have a low and sometimes negative radiation balance. A tiny amount of solar heat reaches the surface. Sometimes it is smaller than the heat radiated back into outer space. The duration of polar night depends on the geographical latitude. For example, the Murmansk region has 59–62 days of polar night [72, 73]. The cold maritime climate is harmful to health. In such conditions, a top priority is the health of children exposed to extreme natural and climatic conditions [73].

Pneumonia progresses peculiarly in the Arctic. Lipid peroxidation (LP) is accompanied by heightened free radical activity and the breach of antioxidant (AO) defense. The LP/AO imbalance impairs pulmonary ventilation and reduces bronchial permeability [68].

During adaptation to living in the Arctic, young males had the worst health scores on day 20 after arrival. Adaptation optimization by reducing the amount and intensity of physical exercise proved insufficient. In only 31.2% of subjects, body functions returned to their initial level by the end of month 2. Moreover, 11.7% of newcomers had symptoms of poor acclimatization. Energy imbalance, intense physical activity, and inadequate carbohydrate-rich diet impeded acclimatization in the Arctic [3].

The Primorsky region has a different type of northern maritime climate, temperate and monsoonal. Winters are dry and cold with clear weather, while springs are long and chilly with temperatures fluctuating. Precipitation is heavy in the warm and humid summer. In some areas, the season accounts for up to 90% of annual precipitation. Most autumns are warm, dry, and cloudless. Southern winds from the Pacific prevail in the summer; northern winds blow in the winter, bringing cold and clear weather from the continent. Heavy summer precipitation and frequent fogs are distinctive features of the climate in the region. The cold summers observed on the coast of the Strait of Tartary are due to the influence of the cold Primorsky current. The hottest summers are on the Prikhanka Plain. The absolute maximum of +41 °C was registered in the Pogranichny district. The average January temperature ranges on the coast from −8 °C to −18 °C. The humidity and winds make the temperature twice as cold. In continental districts, where the climate is drier, the temperature may reach −38 °C, and the winds are weaker than on the coast.

The warmest winters are in the south of the Khasansky district and on the east coast, while the coldest ones are in the mountainous areas of the central and northern part of the region. The absolute minimum of −54 °C recorded in the Krasnoarmeisky district near the village of Glubinnoye is due to a temperature inversion in the intermountain basin. The coldest months are December, January, and February. Annual precipitation is 600–900 mm [74].

According to Veremchuk and Chelnokova [95], living in the Primorye region is stressful to human health because of a high risk of environment-related diseases (respiratory conditions, urinary system disorders, cancer) [75].

The environmental and hygienic conditions of the Primorye region have been shown to influence acclimatization and morbidity among young males. The adverse effects include severe compensatory responses from the cardiovascular, respiratory, vegetative, immune, and pituitary-adrenal systems.

The poor health of young males and their reduced potential for acclimatization affect the capacity of the cardiovascular, respiratory, autonomic, and immune systems. These factors influence their performance in the Primorye region.

Non-locals have a high incidence of community-acquired pneumonia and suffer from severe clinical symptoms once they have contracted it. High intoxication levels, respiratory tract lesions, and many complications accompany the disease. People relocating to the region in the autumn have difficulty adapting to the autumn-winter outbreak of community-acquired pneumonia. Their illness is usually severe, with pronounced symptoms. Those arriving in the spring, before the seasonal epidemic, have time to adapt to the new climatic and geographical conditions. If they contract pneumonia, it has a more benign course in them than in those who have relocated in the spring [76].

Kamchatka is Russia's second-largest peninsula, located on the eastern edge of Eurasia and covering about 15.0% of the Far East. Its unique climate resembles neither the monsoonal climate of the Primorye region nor the oceanic one of the Kuril and Komandor islands. Having no counterparts in Russia, it is not typical for its geographical latitude. The climatic variance observed in the region is a product of cyclones, large water areas, considerable thermobaric contrasts, and complex terrain.

Four climatic regions are distinguished in the Kamchatka region: the north, the Kamchatka River valley, the east coast, and the west coast.

The south-east coast has a temperate maritime climate. Seasonal winds are often gusty; winters are mild but long; summers are chilly. Rapid transit of cyclones causes heavy precipitation and weather variability.

The average annual temperature in the coldest month, February, is 7–8 °C. The average monthly wind speed is 6–10 m/s. Extensive snow events, hurricane winds, and abrupt drops in atmospheric pressure are frequent. Summers are cool, cloudy, and humid. The average monthly temperature in July and August does not exceed +13 to 14.5 °C. An intense breeze circulation causes cold, wet winds, a low cloud cover, drizzle, and fog.

The region lies within a zone of intense cyclone activity. Cyclones move at great speed, and the faster they do, the more dramatically the weather changes. More than a hundred cyclones pass through Kamchatka and its environs each year.

During 24 h, the air temperature may change by as much as 8–10°. Winds have a distinct seasonal pattern: north and north-west winds blow in cold weather; south, in warm. In the winter, the wind speed may exceed 40 m/s. The annual number of days with winds higher than 15 m/s is 134. The average annual cloudiness is about 5 okta. The winter lasts from November to March with 180–210 days of snow cover of an average thickness of 112.0 cm (the maximum snow cover is 2.0 m). Precipitation in the summer, albeit less heavy than in the winter, is from 350 to 500 mm. Winter temperatures are often positive (from 3 to 5 °C) due to southern cyclones.

The Kamchatka weather service distinguishes five weather types. Each is identified based on the atmospheric pressure, temperature, humidity, the gradients of these three quantities, wind speed and direction, and precipitation. The first type is anticyclone decay and the approach of a warm atmospheric front; the second, the emergence of a low-pressure area (a cyclone or trough); the third, the approach of a cold atmospheric front; the fourth, a cold atmospheric front and a high-pressure area (ridges). The mixed type is a combination of warm and cold fronts (their passage and approach).

The first and second types of weather (very fair and fair) have tonic and hypotensive meteoropathy effects. The third and fourth types (foul and very foul) have spastic and hypoxic meteoropathy effects. The mixed type produces a combination of several impacts [77, 78].

The most common category of diseases among 25–50-year-olds in Kamchatka was respiratory diseases, followed by the conditions of the musculoskeletal system and connective tissue and those of the eye and adnexa. The fourth most common category was infectious and parasitic diseases; the fifth, digestive diseases; the sixth, injuries and other health conditions due to an external cause; the seventh, diseases of the urogenital system. Together, they accounted for 77.1% of all diseases, including newly diagnosed cases. Thus, seven classes of diseases were the most widespread [79].

The incidence rate of respiratory diseases was 2.8–3.2 times that of the categories ranked second and third, and 11.4 that of the group of conditions ranked tenth. The incidence rates of the second and third most common categories of diseases did not differ statistically ($p = 0.176$). Still, the incidence rate of these two categories (diseases of the musculoskeletal system and connective tissue and those of the eye and adnexa) was statistically higher than that of other categories ($p = 0.000$). The incidence rates of infectious and parasitic diseases, digestive diseases, injuries, and other health conditions due to an external cause, diseases of the urogenital system and skin and subcutaneous tissue diseases did not differ significantly either. Yet, there was a 1.5-fold difference between the incidence rates of circulatory system disorders and diseases of the ear and the mastoid process ($p = 0.029$).

The Bering Sea coast is a tundra zone with a maritime climate with frequent monsoons, that is, ocean and land influences alternate there as seasons change. A principal climatic feature of the region is strong winds. Cold continental currents cancel out the warming effect of the sea in the winter, while cold air masses moving from the sea increase cloudiness in the summer when fogs and precipitation are frequent and solar radiation is reduced.

Winds grow stronger between November and March. Weak winds, the weakest being south-west ones, blow very rarely – only 10% of the time. North-west winds are frequent in cold months; south-east, in warm. The winds of these two directions are the strongest, reaching speeds of over 15 m/s. Most storm and hurricane winds are from the east. They may be as high as 40–50 m/s in the winter, accompanied by severe snowstorms, and 30–35 m/s in summer, causing storm surges in the bay.

The winter lasts 7 months, from October to April. Strong winds and intense snowstorms are frequent in the season. The average monthly wind speed is 8–10 m/s, and the number of days of snowstorms ranges from 60 to 80. During periodic thaws, temperatures reach 5 °C. The number of cloudy days is high.

Spring lasts only 1 month, May. Snow usually melts in the second half of the month.

The cold and wet summer is from June to the second decade of August. Fog is frequent during the season.

The autumn lasts from the third decade of August to the end of September. Overcast, cold, and damp weather is typical, along with mixed precipitation and light winds.

Snow cover appears in the second decade of September.

The average annual air temperature ranges from −8 to 9 °C. January is the coldest month; July, the warmest. The average temperature in January is −22 to 23 °C; in July, +12 to 13 °C. The record low is −47.0 °C; the record high, +28 °C.

Mean daily temperatures usually go below zero at the end of May and at the end of September. Night frosts are possible as early as August. Moisture carried from the Pacific Ocean strongly influences precipitation, whose total annual value is 250–300 mm. Most precipitation occurs from July to August and in October, and the least falls in March. August has the heaviest rainfall of 46 mm. In some years, the annual precipitation exceeds 500 mm. Solid precipitation occurs during 7–8 months. The difference between the precipitation and the water stored in the snow cover is substantial due to strong winter winds. Fresh snow is redistributed between open and sheltered areas, much of it blown into the sea.

The water surface temperature in the Anadyr Liman is about +10 °C in the summer. Sometimes it reaches as high as +16 °C. The average ice thickness at the end of the winter is about 1 m. Summer navigation in Anadyr usually starts at the end of June and ends in mid-October.

Anadyr is a harsh place to live in from the bioclimatological point of view. It is particularly so for people suffering from diseases of the cardiovascular and nervous systems and bronchopulmonary disorders.

Sharp variations in the pressure and temperature, strong winds, low oxygen content, and frequent cloudiness affect people suffering from meteoropathy [77].

The most common category of diseases among 25–50-year-olds working together in Chukotka was respiratory diseases. Digestive diseases ranked second. The third most widespread category was diseases of the musculoskeletal system and connective tissue. The fourth was diseases of the circulatory system; the fifth, injuries and other health conditions due to an external cause; the sixth, disorders of the skin and subcutaneous tissue; the seventh, diseases of the genitourinary system. These seven most frequent disease categories accounted for 92% of the total morbidity.

The incidence of respiratory diseases was statistically higher than that of other categories by a factor of 1.9 (diseases of the digestive system) and 1.8 (infectious and parasitic diseases). The incidence of digestive diseases was statistically higher than that of all other categories ($p = 0.000$). Diseases of the musculoskeletal system and connective tissue also had a higher incidence than other conditions ($p = 0.003$ for disorders of the circulatory system, $p = 0.000$ for other classes). Another group of disease categories where incidence rates did not differ statistically included circulatory system diseases, injuries, and other health conditions due to an external cause, and skin and subcutaneous tissue diseases. The incidence rate of urogenital system diseases was above that of other categories. The remaining categories – diseases of the eye and adnexa, diseases of the ear and the mastoid process, and infectious and parasitic diseases – did not differ statistically in terms of incidence.

The Magadan region is at the north-eastern edge of Asia. A vast area, it has a complex and varied terrain. It is bounded in the south-east by the cold waters of the Sea of Okhotsk, part of the Pacific Ocean basin. The region lies within the harsh subarctic zones of tundra and wooded tundra, known for high humidity, cold summers, and snowy winters.

Two types of climate are distinguished in this zone based on thermal conditions in the winter:

1. A continental climate with bitter winters (mainly in the vast continental part of the region).
2. A temperate continental and maritime climate with moderately severe winters. Almost the entire Magadan region is located in the permafrost zone.

The average annual air temperature is below zero everywhere in the region, varying from −2 to 3 °C on the coast of the Sea of Okhotsk to −11 to 13 °C in the continental areas.

The frost-free period is relatively long because of the influence of the sea. Temperatures rarely drop below zero in the second half of June; frosts return at the end of September on most of the coast.

The frost-free period lasts, on average, more than 100 days. In some years, negative temperatures are observed in continental areas throughout the entire year.

Wind speeds vary considerably, increasing toward the coastline. While the average annual wind speed is 2–3 m/s in continental areas, it may reach 7 m/s on the coast. Wind speeds rarely exceed 20 m/s in the central parts of the region and 40 m/s at the Sea of Okhotsk coast.

Winter lasts from 6 months in the south to seven and a half months in the north. The snow cover usually settles by mid-October. In some years, it appears as early as the beginning of the month to stay until May. Strong winds accompanied by long snowstorms are frequent in the winter. The average number of days when snowstorms sweep across the coast is 30–40 or even 60 in some places (Shelikhovo, Spafaryeva, Magadan, Brokhovo). In the central parts of the region, where wind speeds are low during the cold season, snowstorms occur 10–20 days a year.

Abrupt changes in the weather, frequent storms, hard frosts, thick fogs, extensive snowstorms, and heavy precipitation take a toll on the performance of organizations in the Magadan region [77].

In Magadan, two disease categories were the most widespread among 20- to 50-year-olds: digestive disorders and respiratory diseases. Their incidence rates did

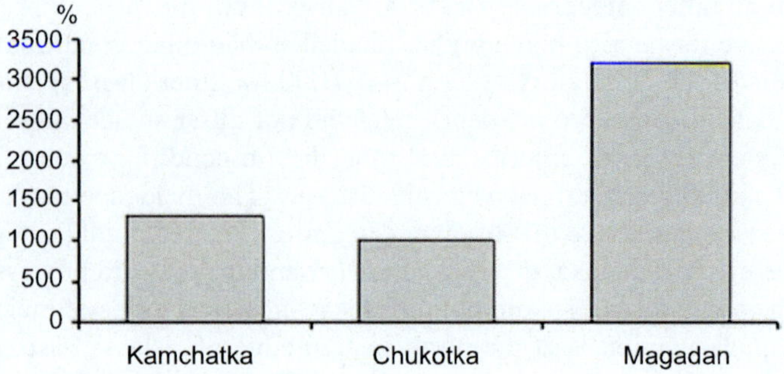

Fig. 5.11 Morbidity in adults working in groups

not differ significantly ($p = 0.423$). Their rates were 1.4 times those of diseases of the musculoskeletal system and connective tissue ($p = 0.000$). Diseases of the eye and adnexa were the third most widespread category, with morbidity lower than in the first two categories by factors of 3 and 2.2. Diseases of the endocrine system, nutritional and metabolic disorders, and diseases of the urogenital system ranked fourth. The morbidity of infectious and parasitic diseases, diseases of the circulatory system, and mental and behavioral disorders did not differ statistically. Neoplasms, ear and mastoid process diseases, and skin and subcutaneous tissue diseases ranked fifth. Overall, 85.8% of the total morbidity was accounted for by six categories of disorders: digestive diseases, respiratory diseases, diseases of the musculoskeletal system and connective tissue, diseases of the eye and adnexa, diseases of the endocrine system, nutritional, and metabolic disorders, and diseases of the urogenital system.

Morbidity was the highest in Magadan, the lowest in Chukotka. In Magadan, it was 2.4 and 2.9 times that in Kamchatka and Chukotka ($p = 0.000$) (Fig. 5.11).

Comparing the incidence rate of the most widespread disease categories shows that respiratory disorders were 1.6 and 1.7 times more prevalent in Magadan than in Kamchatka and Chukotka (Fig. 5.12). Diseases of the musculoskeletal system and connective tissue occurred in the city 3.2 and 4.1 times as often as the two other areas (Fig. 5.13). Morbidity in Kamchatka and Chukotka did not differ ($p = 0.121$).

Diseases of the eye and adnexa occurred in Magadan 1.6 times as frequently in Kamchatka and 19.2 times as often as in Chukotka (Fig. 5.14). Yet, in Kamchatka, it was 11.7 times as prevalent as in Chukotka.

The incidence of infectious and parasitic diseases in Kamchatka and Magadan did not differ statistically ($p = 0.643$). In Chukotka, however, the incidence was lower than in these two areas by factors of 23.1 and 24.2, respectively (Fig. 5.15).

Residents of Magadan and Chukotka often suffered from digestive diseases. In Magadan, the morbidity of this category of disorders was 3.2 and 9.3 times that in Chukotka and Kamchatka (Fig. 5.16).

The rates of injuries and other health conditions due to an external cause did not differ statistically ($p = 0.17$ for Kamchatka and Chukotka; $p = 0.411$ for Kamchatka and Magadan; $p = 0.403$ for Chukotka and Magadan) (Fig. 5.17).

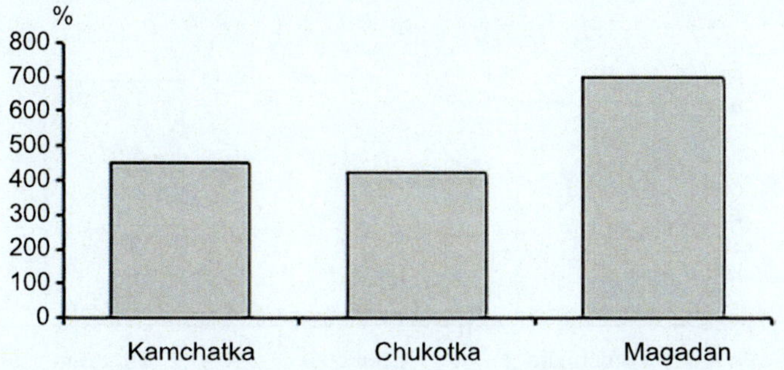

Fig. 5.12 The incidence of respiratory diseases

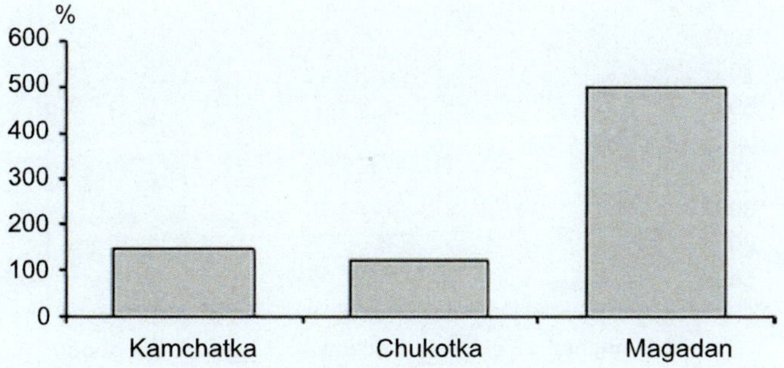

Fig. 5.13 The morbidity of diseases of the musculoskeletal system and connective tissue

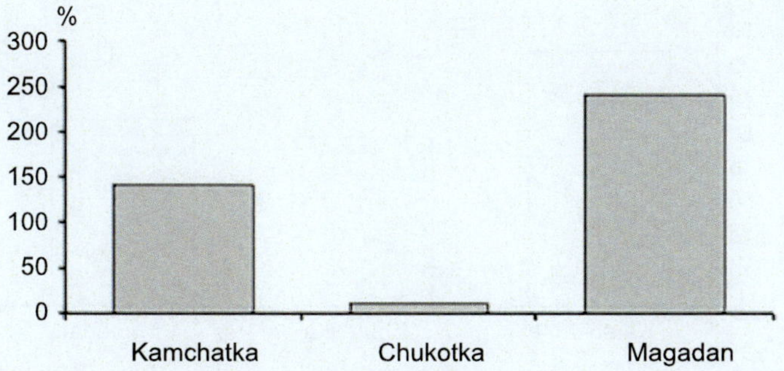

Fig. 5.14 The incidence of diseases of the eye and adnexa

The incidence of diseases of the urogenital system was the highest in Magadan: 2.6 times that in Kamchatka and 5.5 times in Chukotka (Fig. 5.18). The incidence in Kamchatka was higher than in Chukotka by a factor of 2.1 ($p = 0.002$).

The incidence of diseases of the skin and subcutaneous tissue in Magadan was 0.67 times that in Chukotka and 0.45 that in Kamchatka (Fig. 5.19). In Kamchatka, the incidence was higher than in Chukotka by a factor of 1.5 ($p = 0.01$).

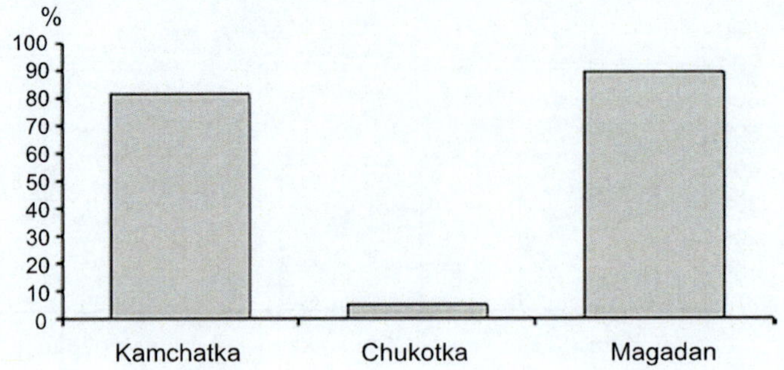

Fig. 5.15 Morbidity of infectious and parasitic diseases

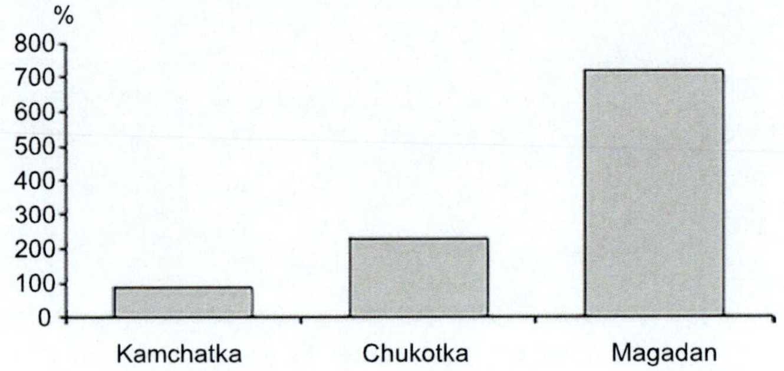

Fig. 5.16 The incidence of digestive diseases

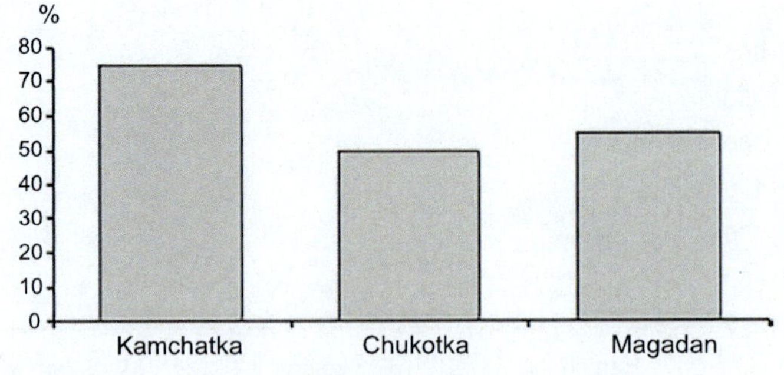

Fig. 5.17 The incidence of injuries and other health conditions due to an external cause

The incidence of diseases of the circulatory system was similar in Kamchatka and Chukotka ($p = 0.224$), and it was significantly higher in Magadan – by factors 1.4 ($p = 0.007$) and 1.7 ($p = 0.000$), respectively (Fig. 5.20).

In the case of diseases of the ear and the mastoid process, the incidence did not differ in Kamchatka and the Magadan region ($p = 0.192$), while it was the lowest in Chukotka – 0.26 times that in Kamchatka ($p = 0.000$) and 0.36 times that in Magadan ($p = 0.000$) (Fig. 5.21).

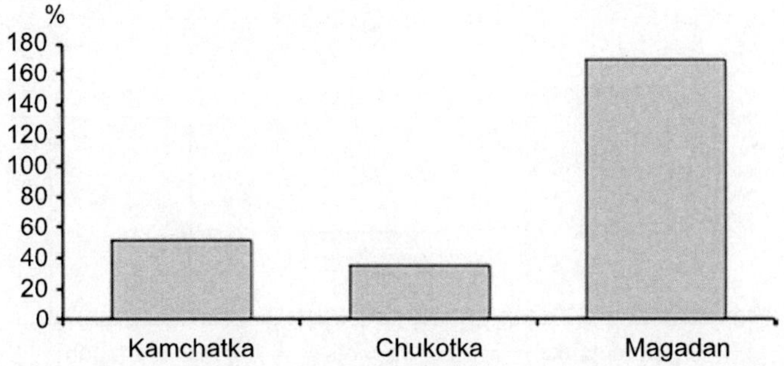

Fig. 5.18 The incidence of diseases of the urogenital system

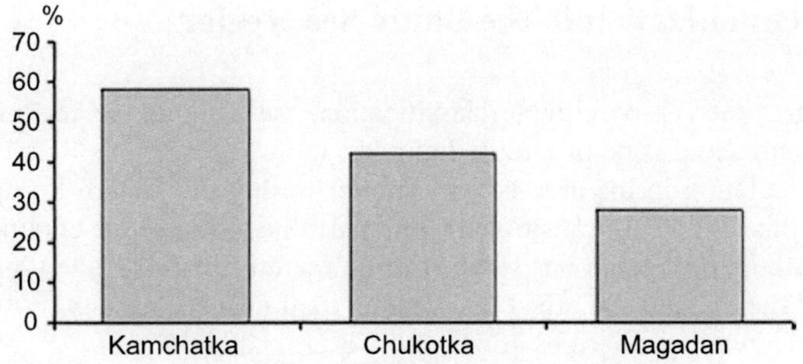

Fig. 5.19 The incidence of diseases of the skin and subcutaneous tissue

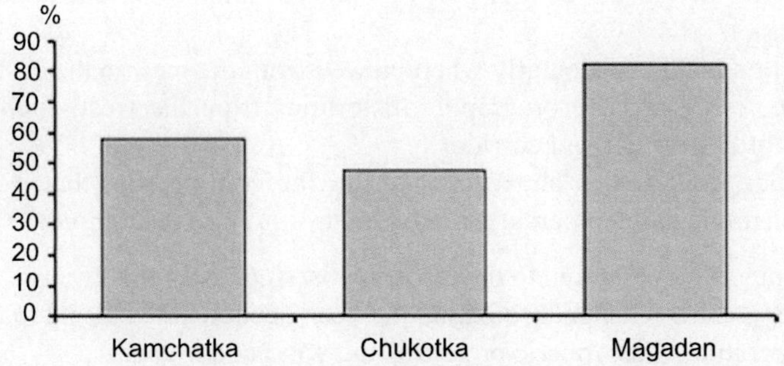

Fig. 5.20 The incidence of diseases of the circulatory system

Sakhalin also has a maritime monsoonal climate that harms human health. The island and the Kurils lie in the middle latitudes monsoonal zone. On Sakhalin, the climate is monsoonal; on the Kurils, oceanic. Both, however, are much harsher than in other areas of the temperate zone [80].

When exposed to the monsoonal climate of Sakhalin, asthma sufferers are afflicted by fluctuations in temperature, air humidity, and pressure as seasons change. The effect of these factors is the strongest in November and March [81].

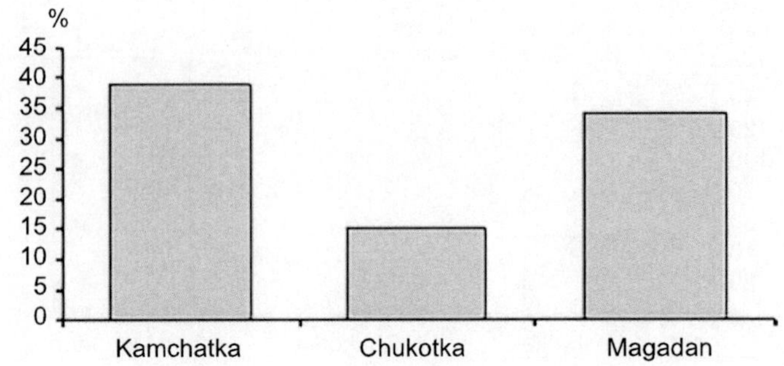

Fig. 5.21 The incidence of diseases of the ear and the mastoid process

5.5 Acclimatization in the Baltic Sea Region

According to the Alisov climate classification, the Kaliningrad region lies in the Atlantic continental zone of middle latitudes, more precisely, its South Baltic subregion. Circulation in the area is very similar to that in Western Europe. Köppen described this climate as transitional from maritime to temperate continental [82].

The Kaliningrad region has some specific features that alter the typical characteristics of the climate and affect the weather in almost all seasons.

These regional peculiarities arise out of several natural factors:

- The region is located between the vast expanse of the Atlantic Ocean and Eurasia, the largest continent on the globe.
- The region lies in the central part of both the middle latitudes and the Baltic Sea coast.
- The terrain is flat, particularly when viewed from the west to the east.
- The area is exposed to orographic distortions from the west, south-west, and partly north-west (a wind corridor).
- The polar front settles above the area in the cold season, that is, the divide between arctic and temperate air masses runs through the region.

Numerous cyclones come to the territory, passing along this front.

Weather patterns change throughout the year, accompanied by sharp fluctuations in air temperature, atmospheric pressure, and wind conditions.

5.5.1 Solar Radiation

The annual radiation balance in the Kaliningrad region varies from 1630 MJ/m^2 (Svetlogorsk) to 1500 MJ/m^2 (Chernyakhovsk). From November to February, this balance is negative. The effective radiation in this period is below that in the summer. Moreover, the absorbed radiation is further reduced by the short day length and the snow cover reflecting radiation.

A thick overcast of clouds appears quite often throughout the year. The cloud cover frequently exceeds five oktas. High air humidity and cloudiness significantly affect the light regime. Clear days are few, not more than 30–33 a year.

Overcast days prevail in the autumn and winter when the sunshine duration is only 12–23% of the possible maximum. Sunshine duration varies from year to year. For example, 1953 had 2082 hours of sunshine, and 1970 only 1538. Cloudiness, dustiness, and other atmospheric circulation factors unrelated to geographic latitude also affect radiation.

As the sky in the Kaliningrad region is often overcast, direct radiation is supplemented by scattered radiation, especially when the sun is low. The area receives only 6% of the annual amount of radiation in the winter; 33% in the spring; 45 in the summer; 16 in the autumn. The Baltic coast and the eastern part of the region get most of the radiation; the central and northern parts, the least.

5.5.2 *Atmospheric Circulation*

Maritime and continental air masses of different origins constantly alternate in the Kaliningrad region. Temperate air masses, prevailing 81% of the time, shape the climate. Maritime air masses of temperate latitudes, which come with cyclones from the Atlantic, bring thaws in the winter, making the cold season less severe. Maritime air is responsible for clear weather replaced by rainfall, low temperatures, gusty winds, and thunderstorms in the summer. East streams carry the continental air of temperate latitudes from the Eurasian continent. This air mass is warm in the summer and cold in the winter. In the summer, temperatures are high, sometimes reaching +28 °C; fogs and weak winds are frequent. In the winter, continental air causes the temperature to drop below –20 °C; easterly winds blow, and the weather is frosty and mainly clear (8).

5.5.3 *Wind Conditions*

For most of the year (173 days), the region is influenced by cyclonic circulation. Anticyclones affect the weather for an average of 133 days. About 175 frontal divides pass through the area per year, causing overcast weather and moderate to high winds.

High wind activity is typical for the region in almost all seasons. From the west to the east and from the south to the north, the proportion of days with 2–20 m/s winds varies from 67 to 81%.

Winds higher than 20 m/s, that is, storms and hurricanes, pose considerable danger. During the 40 years of the study, such winds blew four to five times a year, making a total of 112. Storm winds prevail in the autumn and winter (91.9%), while only 8.1% occur in the spring and summer. One hundred three out of the 122 storm events happened in the autumn (52) and winter (51). Only eight events took place in the spring. In the summer, they are exceptionally rare. The breeze circulation in the region is typical for coastal areas, breezes being a product of anticyclones and calm weather. The physical prerequisites are not always present in the Kaliningrad region: these winds require a marked temperature contrast between the land and the sea surface.

In the daytime, breezes can be as high as 5–7 m/s and blow tens of kilometers inland. Strong winds of over 7 m/s are dangerous for people suffering from cardiovascular diseases and asthma [83].

5.5.4 *Humidity*

The water vapor pressure is substantial in the Kaliningrad region due to the transport of humid air from the ocean. It is the highest in July (15–16 hPa) and February (4.0–4.8 hPa). The annual values of water vapor pressure vary slightly across the region, between 8 and 9 hPa. Relative humidity in the Kaliningrad region is very high, the average annual reaching 81–82%. In the winter and autumn, the relative humidity rises to 85–87%. The minimum values, 71–73%, are observed at the beginning of the summer (in May–June). Exceptions are coastal areas, where the relative humidity is constantly high (78–80%). It drops as low as 55% only in the east of the region on early summer afternoons [84].

5.5.5 *Precipitation*

The Kaliningrad region is an area of high humidity. Intense cyclone activity results in the average annual precipitation of 750–800 mm. The warm period from April to October accounts for 60–65% of the annual precipitation, of which there are 178–183 days. Most of the precipitation is rainfall from heavy clouds carried by the cold or warm fronts of Atlantic cyclones. Rain and snow mixed is frequent in the cold seasons when air temperatures are close to 0 °C. Solid precipitation (snow, snow pellets) accounts for only 8–10% of the total. A stable snow cover forms very late, on December 27–31, and stays until mid-March. Winters without a stable snow cover account for 47% of all winters in Kaliningrad. Rainfall is frequent, occurring almost every second summer day (13–16 days per month). However, the intensity of precipitation is very different in the spring and the summer. In the latter case, rains usually turn into short downpours accompanied by thunderstorms and strong winds. There are 50–60 h of such precipitation per month in the summer and 120–160 h in the winter. Continuous fogs are frequent in the cold season. They are due to the warm air brought by south-west winds. There are 60 days with fog in the Neman River delta and fewer in the south-east and south-west. The average duration of fogs is 55 h; it is the highest in November–December, the spring, and March, amounting to 30 h a month in Kaliningrad [84].

Table 5.8 Prevalence of diseases and incidence in the treatment group exposed to the influence of the Baltic Sea

Category of disease	Incidence per 1000 population, ‰
1. Infectious and parasitic diseases	65.8 ± 10.8
6. Nervous system diseases	87.1 ± 9.9
7. Diseases of the eye and adnexa	38.6 ± 13.7
8. Diseases of the ear	13.6 ± 2.6
10. Respiratory diseases	1655.8 ± 267.1
11. Digestive diseases	60.2 ± 7.4
12. Skin and subcutaneous tissue diseases	265.8 ± 24.6
13. Diseases of the musculoskeletal system and connective tissue	239.4 ± 19.2
14. Diseases of the urogenital system	28.2 ± 5.5
19. Injuries, poisoning, and other health conditions due to an external cause	182.8 ± 21.1
Other diseases entities	24.5 ± 3.1
Total for all categories	2655.9 ± 301.5

Since his type of climate adversely affects human health, the Kaliningrad weather and climate harmed the health of young males during job adaptation [56]. An analysis of incidence among young males shows that the most prevalent conditions belonged to the following categories: respiratory diseases (63.1%); skin and subcutaneous tissue diseases (10.3%); diseases of the musculoskeletal system and connective tissue (9.5%); injuries, poisonings, and other health conditions due to an external cause (6.8%). Together, these categories accounted for 89.7% of all cases.

The incidence of respiratory diseases was 6.2 times that of the second-ranked category (skin and subcutaneous tissue diseases); 6.9 times of the third-ranked category (musculoskeletal and connective tissue diseases); 9.1 times that of the fourth-ranked class (injuries, poisoning, and other health conditions due to an external cause) (Table 5.8).

Incidence among people who had spent 1 year in the region was 1.6 times that among 2-year residents; twice, among 3-year residents; 2.7-times, among 4-year residents (Fig. 5.22).

The incidence of the main categories of diseases among first-year residents was higher than among individuals who had spent 2, 3, or 4 years in the region:

- By factors of 1.8, 2.6, and 4 for respiratory diseases.
- By factors of 1.2, 1.5, and 2.1 for skin and subcutaneous tissue disorders.
- By factors 1.7, 1.4, and 1.5 for diseases of the skeletal and muscular systems and the connective tissue.

Analysis of how incidence depends on the period spent in the region reveals the following patterns. The absolute decrease in infectious and parasitic diseases was 14.1 per mille, with a rate of annual decline of 21.3% (Table 5.9).

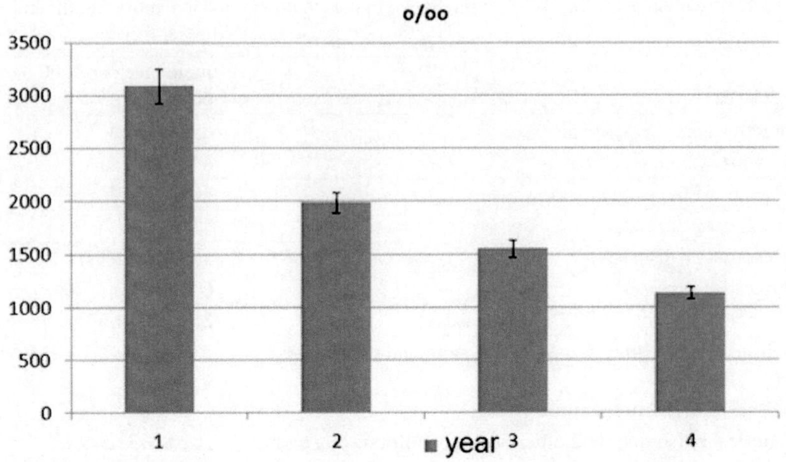

Fig. 5.22 Incidence depending on the years spent in the region, ‰

Table 5.9 Incidence in young males exposed to the influence of the Baltic Sea for different periods, ‰

Disease category	1 year	2 years	3 years	4 years
1. Infectious and parasitic diseases	69.2 ± 12.7	46.8 ± 5.6	56.8 ± 2.5	26.8 ± 6.47
6. Nervous system diseases	61.1 ± 8.3	99.5 ± 20.9	83.9 ± 7.7	80.5 ± 16.1
7. Diseases of the eye and adnexa	66.3 ± 15.4	25.1 ± 3.2	15.5 ± 4.7	13.4 ± 3.2
8. Diseases of the ear	19.5 ± 4.7	8.0 ± 3.2	15.4 ± 3.1	3.4 ± 1.7
10. Respiratory diseases	2056.9 ± 131.4	1147.9 ± 93.1	777.8 ± 103.9	513.2 ± 94.8
11. Digestive diseases	53.7 ± 5.5	54.8 ± 8.3	37.9 ± 9.3	40.1 ± 16.3
12. Skin and subcutaneous tissue diseases	275.6 ± 21.1	227.3 ± 24.1	183.2 ± 17.5	130.6 ± 30.7
13. Diseases of the musculoskeletal system and connective tissue	278.6 ± 23.4	166.4 ± 32.2	195.3 ± 26.9	187.8 ± 35.4
14. Diseases of the urogenital system	25.1 ± 3.8	16.6 ± 5.7	37.8 ± 13.1	23.5 ± 1.5
19. Injuries, poisoning, and other health conditions due to an external cause	177.2 ± 17.7	178.8 ± 26.8	134.9 ± 23.6	100.2 ± 35.8
Other diseases entities	18.1 ± 5.3	21.0 ± 6.6	13.2 ± 0.2	13.4 ± 0.1
Total for all categories	3084.7 ± 192.3	1985.3 ± 198.7	1551.7 ± 197.9	1133.1 ± 238.5

The absolute increase in diseases of the nervous system was 6.5 per mille; the annual increment, 14.4%.

The absolute decrease in diseases of the eye and adnexa was 17.6 per mille; the rate of annual decline, 37.9%.

The absolute decrease in diseases of the ear was 5.4 per mille; the rate of annual decline, 14.8%.

The absolute decrease in respiratory diseases was 514.5 per mille; the rate of annual decline, 36.8%.

The absolute decrease in digestive diseases was 4.5 per mille; the rate of annual decline, 7.6%.

The absolute decrease in skin and subcutaneous tissue diseases was 48.5 per mille; the rate of annual decline, 21.9%.

The absolute decrease in diseases of the skeletal and muscular systems and connective tissue was 30.3 per mille; the rate of annual decline, 8.9%.

The absolute decrease in diseases of the urogenital system was 30.3 per mille; the rate of annual decline, 18.7%.

The absolute decrease in injuries, poisoning, and other health conditions due to an external cause was 25.7 per mille; the rate of annual decline, 16.5%.

A comparative analysis of incidence among permanent residents of the Kaliningrad region (adapted individuals) and newcomers from other Russian territories shows that 90.5% of the latter originated from areas with different climate and weather conditions. Most of them had come from the temperate zone (80%), while 10% had relocated from the Arctic and subtropical climate.

Incidence rates differed statistically. Among newcomers, the incidence was 1.5 times that among locals (Table 5.10). The difference was statistically significant for five categories of diseases. The incidence of respiratory disorders was 1.5 times higher among newcomers than among permanent residents; of skin and subcutaneous tissue diseases, 1.4 times; of diseases of the musculoskeletal system and connective tissue, 1.8 times; of injuries, poisoning, and other health conditions due to an external cause, 1.5 times; of infectious and parasitic diseases, 2.3 times.

Table 5.10 Comparative incidence among people adapted and unadapted to the maritime climate, ‰

Disease category	Locals	Newcomers	$p =$
1. Infectious and parasitic diseases	32.9 ± 4.1	76.4 ± 11.4	0.014
6. Nervous system diseases	89.2 ± 8.5	88.2 ± 8.3	0.943
7. Diseases of the eye and adnexa	39.1 ± 11.2	48.8 ± 8.4	0.512
8. Diseases of the ear	18.5 ± 4.7	7.5 ± 3.3	0.063
10. Respiratory diseases	1287.5 ± 132.8	1921.4 ± 172.3	0.023
11. Digestive diseases	62.6 ± 3.7	67.5 ± 3.9	0.447
12. Skin and subcutaneous tissue diseases	213.3 ± 19.0	307.5 ± 20.8	0.006
13. Diseases of the musculoskeletal system and connective tissue	151.1 ± 10.3	284.9 ± 18.2	0.000
14. Diseases of the urogenital system	24.7 ± 5.8	35.5 ± 4.2	0.148
19. Injuries, poisoning, and other health conditions due to an external cause	145.6 ± 11.0	227.4 ± 6.8	0.000
Other diseases entities	21.4 ± 4.9	22.5 ± 4.7	0.889
Total for all categories	2086.1 ± 153.2	3087.6 ± 271.5	0.019

Table 5.11 Comparative incidence of respiratory diseases among locals and newcomers, ‰

Disease	Locals	Newcomers	$p =$
Acute sinusitis	44.3 ± 8.3	77.5 ± 10.1	0.043
Acute tonsillitis	103.2 ± 12.1	157.1 ± 17.0	0.048
Acute URI an influenza	1051.9 ± 51.3	1534.5 ± 106.6	0.004
CAP	54.9 ± 5.1	80.1 ± 8.3	0.05
Acute LRTI	33.1 ± 7.9	72.2 ± 12.2	0.044

Table 5.12 The incidence among people who had spent in the region less than a year and locals, ‰

Diseases	Locals	Newcomers	$p =$
1. Infectious and parasitic diseases	38.5 ± 18.1	80.4 ± 20.7	0.206
6. Nervous system diseases	62.6 ± 13.4	63.9 ± 10.2	0.941
7. Diseases of the eye and adnexa	76.7 ± 16.4	68.8 ± 14.8	0.748
8. Diseases of the ear	38.2 ± 13.1	12.1 ± 8.3	0.086
10. Respiratory diseases	1736.1 ± 123.9	2406.3 ± 169.9	0.012
11. Digestive diseases	67.3 ± 10.5	55.4 ± 10.9	0.77
12. Skin and subcutaneous tissue diseases	242.1 ± 26.3	325.3 ± 20.1	0.017
13. Diseases of the musculoskeletal system and connective tissue	197.6 ± 31.0	338.2 ± 40.6	0.028
14. Diseases of the urogenital system	19.6 ± 6.0	32.2 ± 9.7	0.375
19. Injuries, poisoning, and other health conditions due to an external cause	139.7 ± 20.8	221.8 ± 24.9	0.039
Other diseases entities	23.9 ± 6.5	19.8 ± 8.2	0.752
Total for all categories	2613.7 ± 278.4	3624.2 ± 262.7	0.02

There were differences at the level of specific disease entities as well. For instance, the incidence of acute upper respiratory infections (URI), acute tonsillitis, and CAP in newcomers was 1.5 times that among permanent residents; of acute sinusitis, 1.7 times; of lower respiratory tract infections (LRTI), 2.2 times (Tables 5.11 and 5.12).

This difference was calculated for the most prevalent categories of diseases. The incidence of respiratory disorders in newcomers was 1.4 times higher than in locals; subcutaneous tissue diseases, 1.3 times; diseases of the musculoskeletal system and connective tissue, 1.7 times; injuries, poisoning, and other health conditions due to an external cause, 1.6 times.

As to specific respiratory diseases, the difference in the incidence of acute sinusitis was 1.9-fold; of acute URI, 1.4-fold (Table 5.13).

In the early stages of acclimatization, the body has a greater need for vitamins and minerals. Table 5.14 demonstrates variations in the mineral status of subjects. The serum copper level was within the reference range. Averaged data obtained 15 days later did not suggest a statistically significant change ($p = 0.442$).

There was a rise in the proportion of individuals whose copper levels dropped within the reference range ($F = 4.234$). The percentage of those whose serum copper levels fell below the reference range also increased, from 10 to 30%.

Table 5.13 Incidence of specific respiratory diseases, ‰

Disease entity	Locals	Newcomers	$p =$
Acute sinusitis	48.8 ± 5.7	94.3 ± 8.0	0.000
Acute tonsillitis	148.1 ± 43.1	193.3 ± 32.1	0.419
Acute URI	1361.0 ± 99.0	1881.0 ± 143.2	0.02
Acute LRTI	62.6 ± 16.2	102.9 ± 16.4	0.132
CAP	120.3 ± 37.0	134.8 ± 4.0	0.571

Table 5.14 Variations in mineral levels over time, absolute values, M ± m

Element (reference range)	Initial	After 15 days	After 45 days
Copper, µg/mL (0.7–1.55)	0.81 ± 0.09	0.73 ± 0.05	0.76 ± 0.04
Zinc, µg/mL (0.5–1.5)	1.0 ± 0.08	0.97 ± 0.08	1.13 ± 0.08
Iron, µmol/L (11.6–31.3)	14.2 ± 1.15	16.84 ± 0.86	16.65 ± 0.93
Calcium, mmol/L (2.15–2.57)	2.61 ± 0.02	2.59 ± 0.01	2.58 ± 0.02
Magnesium, mmol/L (0.66–1.07)	0.995 ± 0.02	0.77 ± 0.009	0.86 ± 0.013
Phosphorus, mmol/L (0.87–1.45)	1.257 ± 0.03	1.25 ± 0.03	1.3 ± 0.03
Potassium, mmol/L (3.6–5.5)	5.19 ± 0.12	5.04 ± 0.1	5.25 ± 0.11
Sodium, mmol/L (135–150)	143.6 ± 0.31	149.6 ± 0.35	150.5 ± 0.48
Chlorine, mmol/L (97–108)	100.2 ± 1.11	100.3 ± 0.4	104.3 ± 0.4

No significant changes in zinc levels occurred. However, at the end of the study, 50% of young males experienced a decrease in zinc levels within the reference range.

Although the level of magnesium in blood serum was within the reference range, it fell over time. In the second test, it was 22.6% ($p = 0.000$) below the initial level; in the third, 13.6% ($p = 0.000$). A decrease was observed in 100 and 90% of subjects, respectively.

The level of serum phosphorus was within the reference limits at all stages. No significant changes occurred over time. Yet, phosphorus levels fell in 66.7% of subjects 15 days into the study. In the end, they were below the initial values in 33.3% of subjects.

The levels of iron, calcium, potassium, sodium, and chloride in blood serum were within the reference range at all stages. No significant changes occurred throughout the study.

Vitamin A levels were above the reference range at all stages (Table 5.15). There were no individuals with low vitamin A levels. Changes throughout the study were statistically insignificant (p 1:2 = 0.265).

Vitamin E levels at the beginning of the study were slightly below the reference range; only 60% of subjects had normal values. Fifteen days into the study, the levels fell by 12.5% ($p = 0.019$) and remained such until the end of observations ($p = 0.049$). A significant decrease in vitamin E occurred in 80% of subjects ($F = 4, 474$).

The second and third blood tests showed that none of the subjects had vitamin B1 levels within the reference range, as evidenced by pyruvic acid content. Overall, the vitamin level was 47.1–47.4% below the upper reference limit.

Table 5.15 Variations in mineral levels over time, absolute values, M ± m

Vitamin (reference range)	Initial =	After 15 days	After 45 days
A, µg/mL (0.3–0.6)	0.83 ± 0.05	0.91 ± 0.05	0.91 ± 0.04
E, µg/mL (8–18)	7.75 ± 0.43	6.78 ± 0.39	6.78 ± 0.53
B_1, µg/mL (7–14)[a]		20.6 ± 1.0[b]	20.63 ± 1.0
B_2, µg/% (10–50)		5.85 ± 0.3	6.08 ± 0.3

[a]Based on the pyruvic acid decomposition product: an increase in pyruvic acid levels means a decrease in those of B_1

[b]Data from two tests

By the end of the study, vitamin B2 levels were close to the initial values. In the second test, the number of subjects with a low level of this vitamin was 60%, and it reached 100% at the final stage: B2 levels fell over time.

Acclimatization engaged natural resistance mechanisms in subjects. The initial content of IgG in blood serum was within the reference range. Yet, a statistically significant decrease in its level, by 27.8%, was observed at the end of the study ($p = 0.039$). The initial IgG level was 15.92 ± 1.02 g/L (the normal range is 8.0–17.0 g/L); after 15 days, 15.46 ± 1.39 g/L; after 45 days, 11.49 ± 1.8 g/L. There was a significant increase in the proportion of individuals with IgG levels below the reference limits: 10% initially; 20% after 15 days; 50% at the end of the study ($F = 2.923$).

Since the level of secretory IgA was below the normal in 40% of subjects, its values in the group were also low (0.685 ± 0.09 g/L; the reference range is 0.9–4.5 g/L). A statistically significant decrease in IgA, by 37.5%, occurred at the end of the study ($p = 0.000$, to 0.428 ± 0.03 g/L). The percentage of subjects with low IgA levels was 80% after 15 days and 100% when the study was completed.

IgM levels were initially below the reference limits (0.54 ± 0.07 g/L, compared to 0.6–3.7 g/L). In other periods, the average values were within the normal range. After 15 and 45 days, the level of IgM increased significantly, and the initial value was exceeded 2.9-fold (1.58 ± 0.22 g/L, $p = 0.000$) and 2.5-fold (1.34 ± 0.16 g/L, $p = 0.000$), respectively. The results of the second and third testing did not differ statistically ($p = 0.383$).

The obtained vitamin and mineral levels suggest a metabolic disturbance at the early stage of acclimatization. The levels of total protein and protein fractions were evaluated to prove this hypothesis. Total protein did not change over time, fluctuating between 73.2 ± 0.6 g/L, 71.8 ± 0.7 g/L ($p_1–p_2 = 0.137$), and 74.8 ± 0.9 g/L. Total protein levels remained unaltered after 15 days in only 36.4% of subjects, while the other 63.6% experienced a decrease within the reference range ($\chi 2 = 14.737$, $p = 0.000$). At the end of the study, protein levels were below the initial values in 40% of subjects ($\chi 2 = 7.265$, $p = 0.007$). In addition, the serum γ-globulin level fell over time. It was 18.7% ($p = 0.001$) after 15 days and 19% ($p = 0.000$) at the end of the study.

These findings suggested that acclimatization adversely affected the production of protective immunoglobulins in the body.

Oxidative stress increased. Subjects who had spent several days in the region had a peroxide concentration in serum indicative of high oxidative stress and an average antioxidant activity of blood serum. At the end of the study, the peroxide level remained unchanged, while the serum antioxidant activity decreased by 11.2%, $p = 0.024$ (from 311.5 ± 11.3 to 276.7 ± 9.5 μmol/L). These values signaled a worsening of the antioxidant status.

Test results indicated a reduction in the proportion of individuals with low and moderate levels of oxidative stress. They also suggested a rise in subjects with a high level of stress (up to 75%).

The antioxidant activity of blood serum decreased. The proportion of individuals with a low (up to 37.5%, 2.5-fold) and average (up to 50%, 1.25-fold) activity grew, while the percentage of subjects with a high activity decreased 3.6-fold (from 45 to 12.5%).

5.6 Health Risk Assessment of Weather and Climate Impacts by Bioclimatic Indices

The assessment was carried out on three climatic zones: arctic (region of Cape Chelyuskin), subarctic (Norilsk region), and temperate continental climate (Krasnoyarsk and Minusinsk regions).

5.6.1 Habitat Conditions in the Arctic

Assessment of the physical environment in the open area showed that only in July and August did the mean monthly temperature have positive values. In June it was close to zero, and in August it fluctuated within zero (Table 5.16).

The highest wind speeds were recorded in January and February. The wind strength reached 4 points and was rated as "moderate." During the spring period (March–May) and in November, the wind strength weakened slightly, approaching the upper limit of the value and was estimated as "weak to moderate."

Relative humidity was highest in May–September. Then it began to drop and in November–April (6 months a year) stayed at the same level. In May–September, the air was assessed as "strongly humid" and in November–April as "moderately humid"; October showed transitional values.

The determination of the integral indicator of the conditions for body cooling showed two features:

1. Throughout the year, the risk of frostbite in unclothed parts of the human body was detected. In March–October (8 months) the risk was assessed as moderate and for 4 months it was assessed as critical. Only in July and August did the risk values reach the lower risk limit, which is assessed as moderate.

Table 5.16 Multiyear monthly indicators of the state of physical factors in an open area, in absolute values

Month											
1	2	3	4	5	6	7	8	9	10	11	12
Daily air temperature, °C											
−26.0 ± 1.1	−24.5 ± 1.0	−22.9 ± 1.2	−16.3 ± 0,4	−8.0 ± 0.5	−0.3 ± 0.2	1.4 ± 0.2	2.1 ± 0.4	−0.05 ± 0.4	−6.0 ± 0.6	−17.0 ± 0.8	−22.3 ± 0.7
Wind speed, m/s											
6.4 ± 0.5	6.5 ± 0.4	5.9 ± 0.4	5.8 ± 0.3	5.7 ± 0.2	5.8 ± 0.2	6.1 ± 0.2	5.9 ± 0.3	6.0 ± 0.3	6.2 ± 0.3	6.0 ± 0.4	5.8 ± 0.3
Relative humidity, %											
81.1 ± 0.5	81.6 ± 0.4	81.6 ± 0.4	81.5 ± 0.5	88.5 ± 0.7	89.2 ± 0.9	90.5 ± 0.9	89.6 ± 1.0	88.5 ± 0.6	85.1 ± 0.3	81.1 ± 0.3	82.1 ± 0.3

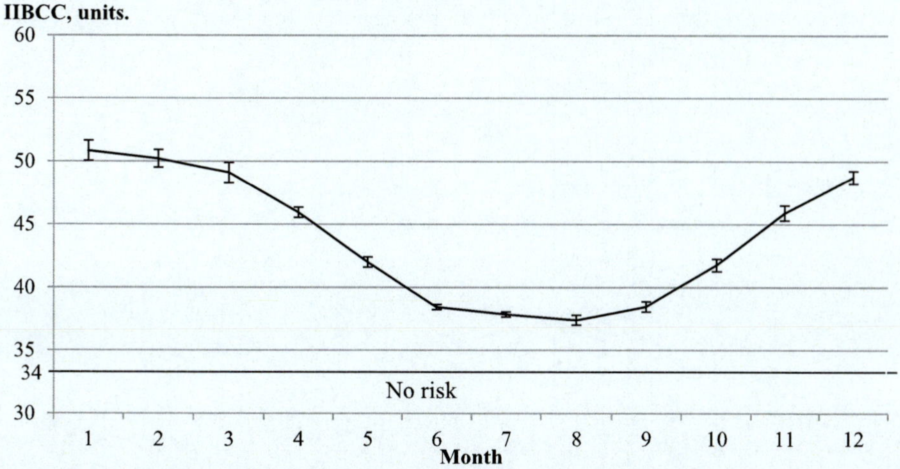

Fig. 5.23 Monthly average values of the IIBCC, units

Table 5.17 Characteristics of weather/climatic parameters according to human heat perception

No. item	Type of heat perception, ET values °C	ET values, °C	Months
1	Cold −12.0 to ≤ −18.0	13.2 ± 0.5 to 16.5 ± 0.5	July–September
2	Very cold −18.0 to ≤ −24.0	–	–
3	Threat of frostbite ˂ −24.0	25.4 ± 0.9 to 47.5 ± 1.3	October–May

2. Indicators of risk criteria were practically equal in July and August; June and September; May and October; April and November; March and December; January and February (Fig. 5.23).

The effective temperature according to A. Missenard also indicated unfavorable weather-climatic conditions (Table 5.17).

ET values in May and October were 1–2 times higher than the "very cold" value, in April and November 1.6 times greater, and in the remaining months 1.9–2.0 times greater.

5.6.2 Habitat in Subarctic and Temperate Continental Climates

In 1990–2020 (the third period of determining climatic norms) in the subarctic belt of the Krasnoyarsk region, the cold period of the year lasted 8–9 months, the warm period 3–4 months (summer and September as a transitional month). In the region of Krasnoyarsk, the cold period of the year lasted 7 months and the warm period 5 months. In the Minusinsk area, the cold period also lasted 7 months and the warm period for 5 months (May–September) (Table 5.18).

Table 5.18 Multi-year monthly air temperatures by observation area, °C

No. item	Observation site	Month, M ± m											
		1	2	3	4	5	6	7	8	9	10	11	12
1	Norilsk	−23.3 ± 1.5	−21.5 ± 1.7	−16.5 ± 1.7	−7.3 ± 0.9	1.6 ± 10	15.3 ± 1.0	19.9 ± 0.9	15.2 ± 0.8	8.2 ± 0.9	−3.6 ± 0.9	−17.8 ± 1.2	−18.0 ± 1.4
2	Krasnoyarsk	−17.3 ± 1.6	−13.9 ± 1.4	−3.9 ± 0.9	4.7 ± 0.5	9.4 ± 0.5	18.2 ± 0.5	18.8 ± 0.3	16.5 ± 0.3	9.4 ± 05	2.4 ± 0.9	−7.5 ± 0.9	−12.6 ± 2.0
3	Minusinsk	−19.9 ± 1.4	−16.6 ± 1.4	−4.5 ± 0.9	5.9 ± 0.5	10.6 ± 0.4	18.9 ± 0.5	19.8 ± 0.4	17.5 ± 0.2	10.4 ± 0.4	2.9 ± 08	−6.4 ± 1.0	−14.3 ± 1.5

Throughout the year, the average daily wind strength in the Norilsk region was estimated at 3 points (weak to moderate): a minimum in June–September, and a maximum (approaching the "moderate" value) in December (Table 5.19). The same wind velocity in the Krasnoyarsk area was assessed for 10 months a year as 2-point "weak wind," and in June–July as 1-point "very weak wind." In the area of Minusinsk during winter months and in July it was estimated at 1 point, while during the rest of the months it was assessed as 2-point "weak wind."

The average minimum temperatures in the region of Norilsk were negative for 8 months of the year, in Krasnoyarsk for 6 months, Minusinsk for 7 months (Table 5.20). The maximum wind was detected in the Norilsk region in May–October when it stood at 6 points (strong wind), and the remaining 6 months at 7 points (from strong to very strong). In January and March, winds of magnitude 8 (very strong) were possible.

In the Krasnoyarsk area, in the summer months the maximum wind was estimated at 3 points (from weak to moderate), and the remaining 9 months at 4 points (moderate); however, winds rating 5 points (moderate to strong) were possible in December and January. In the Minusinsk area, it was windier than in the same temperate climate conditions (Krasnoyarsk): for 4 months (June–September) the wind was estimated at 4 points (moderate), and the remaining 8 months at 5 points (from moderate to strong). In May, wind speeds of up to 6 points (strong) were possible (Table 5.21).

In terms of the average daily temperature and wind, WCI showed a longer period of weather severity in the region of Norilsk: in November (WCI = −9.2 ± 1.5 °C) and April (WCI = −11.8 ± 1.5 °C) the sense of discomfort and coolness was as strong as in March. The same effect was not excluded in October: Q75 equaled 12.1 °C (Fig. 5.24). In November–February (except for January), the WHI was assessed as "very cold, overcooling of the skin surface." The maximum values of Q75 did not go beyond this evaluation criterion. In January, the WCI almost corresponded to the value estimated as "extremely cold, exposed parts of the human body can become supercooled in 10 minutes." Q75 reached 38.1 °C. During the warm period of the year, the Q25 values did not exceed the 10.3 °C limit.

With the combined influence of the minimum temperature and maximum wind in May and October, the severity of the weather was assessed as "discomfort, coolness," in April as "very cold, hypothermia of the skin surface," and 5 months a year as "extremely cold, exposed parts of the body can become supercooled in 10 minutes."

In the Krasnoyarsk area, according to the WCI values, the duration of the cold period of the year was prolonged by 1 month (November was added). According to the average indices, they were within the criterion of "discomfort, coolness"; Q75 did not exceed the upper value of the reference margin. Under the influence of minimum temperature and maximum wind for 5 months of the year, there were conditions assessed as "discomfort, coolness." In January, more severe conditions were possible with a Q25 value of −37.5 °C (Fig. 5.25).

The index in the Minusinsk city area also testified to a possible longer negative period of cold influence on the body: up to 4 months Q75 in October reached

Table 5.19 Multi-year monthly air velocities by observation site, m/s

No. item	Observation site	Month, M ± m											
		1	2	3	4	5	6	7	8	9	10	11	12
1	Norilsk	5.0 ± 0.3	4.6 ± 0.4	4.8 ± 0.4	4.8 ± 0.2	4.2 ± 0.1	4.2 ± 0.1	3.6 ± 0.05	3.6 ± 0.1	3.6 ± 0.2	4.3 ± 0.3	4.4 ± 0.3	5.4 ± 0.2
2	Krasnoyarsk	1.8 ± 0.2	1.9 ± 0.08	2.1 ± 0.1	2.3 ± 0.1	2.1 ± 0.07	1.6 ± 0.07	1.4 ± 0.05	1.5 ± 0.09	1.7 ± 0,1	2.0 ± 0.16	2.0 ± 0.13	1.9 ± 005
3	Minusinsk	1.1 ± 0.1	1.2 ± 0.1	2.0 ± 0.1	2.5 ± 0.06	2.5 ± 0.09	1.8 ± 0.1	1.6 ± 0.07	1.4 ± 0.07	1.6 ± 0.1	2.0 ± 0.1	2.1 ± 0.2	1.4 ± 0.1

Table 5.20 Multi-year minimum monthly air temperatures by observation site, °C

No. item	Observation site	Month, M ± m											
		1	2	3	4	5	6	7	8	9	10	11	12
1	Norilsk	−30.5 ± 1.6	−30.5 ± 1.7	−20.4 ± 2.0	−13.5 ± 0.9	−5.4 ± 0.8	6.5 ± 0.6	11.2 ± 0.6	7.8 ± 0.6	2.2 ± 0.6	−8.9 ± 1.2	−24.8 ± 1.1	−25.3 ± 1.5
2	Krasno-yarsk	−20.9 ± 1.5	−20.8 ± 1.3	−8.8 ± 0.8	−0.3 ± 0.4	4.0 ± 0.4	11.7 ± 0.3	13.6 ± 0.3	11.5 ± 0.3	4.7 ± 0.5	−1.1 ± 0.9	−13.5 ± 1.2	−13.5 ± 1.2
3	Minusinsk	−25.5 ± 1.9	−23 ± 1.4	−10.6 ± 1.0	−1.2 ± 0.4±	2.9 ± 0.3	10.6 ± 0.3	13.0 ± 0.3	10.7 ± 0.3	4.0 ± 0.4	−2.2 ± 0.5	−10.7 ± 1.1	−19.4 ± 1.5

Table 5.21 Multi-year maximum monthly air velocities by observation site, m/s

No. item	Observation site	Month, M ± m											
		1	2	3	4	5	6	7	8	9	10	11	12
1	Norilsk	16.3 ± 0.8	14.1 ± 0.7	16.9 ± 1.0	14.0 ± 0.9	12.8 ± 0.5	12.9 ± 0.9	10.9 ± 1.0	11.2 ± 0.9	12.2 ± 1.4	13.6 ± 1.0	14.4 ± 0.9	15.4 ± 0.6
2	Krasnoyarsk	7.3 ± 0.6	7.3 ± 0.2	6.5 ± 0.2	6.7 ± 0.3	6.9 ± 0.7	5.1 ± 0.4	4.8 ± 0.2	5.3 ± 0.5	6.0 ± 0.4	6.9 ± 0.4	7.0 ± 0.3	79 ± 0.2
3	Minusinsk	6.6 ± 1.0	8.4 ± 0.3	9.0 ± 0.4	10.2 ± 0.3	10.7 ± 0.6	7.3 ± 0.4	7.3 ± 0.4	7.3 ± 0.5	7.5 ± 0.5	9.3 ± 0.5	10.1 ± 0.5	8.5 ± 0.5

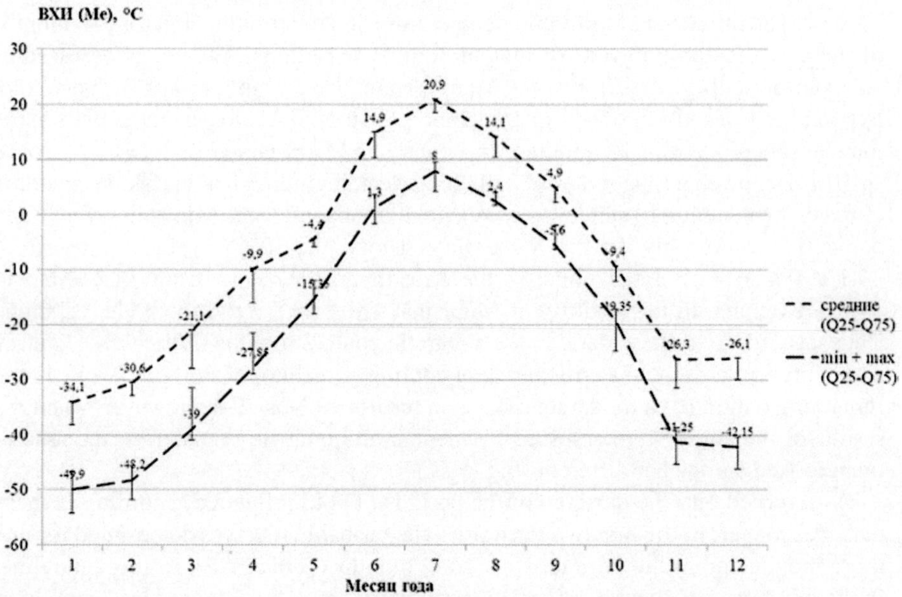

Fig. 5.24 Characteristics of the wind-chill index in Norilsk at various combinations of temperature and wind speed, °C

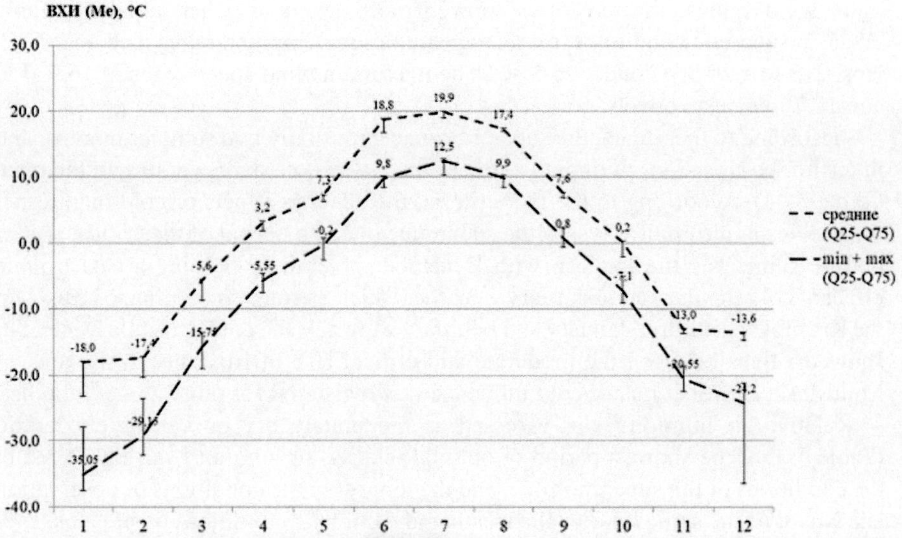

Fig. 5.25 Characteristics of the wind-chill index in Krasnoyarsk conditions at various combinations of temperature and wind speed, °C

13.2 °C. The influence of minimum temperature and maximum wind for 5–6 months of the year created unfavorable meteorological sensations. They were assessed as "discomfort, chilly" in October and March (were also possible in April), "very cold, hypothermia of skin surface" in December, "extremely cold, exposed body parts may be supercooled in 10 minutes" in January and February (Fig. 5.26).

IIBCC showed a longer duration of cold effect, including health risk: in subarctic climate for 8 months (including May), in Krasnoyarsk and Minusinsk for 6 and 5 months accordingly. In the Subarctic territory, the IIBCC values varied from 39.1 ± 0.4 to 48.7 ± 0.7 units, in the Krasnoyarsk territory from 38.2 ± 0.5 to 44.8 ± 1.0 units, in the territory of Minusinsk from 38.8 ± 0.5 to 43.9 ± 1.2 units. Thus, according to these data, in the subarctic zone, 2 months of the year (January and February) presented a critical risk of cooling (frostbite) of the body, while in the remaining 6 months, a moderate risk, as in territories Nos. 2–3. However, when the results of the study were assessed by median and quartile boundaries, the assessment looked somewhat different (Fig. 5.27).

As it turned out, the duration of the periods of cold influence according to these data was longer: in the first two territories, the probable risk was determined for one more month, and in the third territory for 2 months (Table 5.22). At the same time, in the first territory, the risk of critical cooling of the body increased by 1 month; in the second territory by 1 month – moderate, in the third territory – moderate and even critical risk.

When determining the IIBCC according to the minimum temperature and maximum wind values, the territories showed more severe weather variables (Table 5.23). In subarctic conditions, the average minimum temperature reached −30.5 °C, Nos. 2–3 to −20.9 °C and −25.5 °C. The maximum wind speed reached 16.9, 7.9, and 10.7 m/s, respectively.

According to these data, the subarctic zone enjoys only two summer months, and in territories Nos. 2–3, during all summer there was no risk of working in the open (Table 5.24). According to the risk criteria, the adverse effects of cold manifested themselves in different ways. In the subarctic zone, the period of the critical risk of body cooling was the longest, with December potentially holding a catastrophic effect of cold. January and February were the riskiest periods, according to IIBCC. In the Krasnoyarsk region, January and February were also the coldest, while December showed values between the moderate and critical risk of frostbite. In the area of Minusinsk, a critical risk of cold influence was registered for three winter months.

Relative air humidity was assessed as moderately dry or moderately humid (Table 5.25). The shortest period of optimal relative air humidity was registered in the conditions of the subarctic zone, and the longest in the conditions of observation site No. 3. At the same time, in the weather and climatic conditions of site Nos. 2–3, April or May showed potentially lower relative humidity of the air dropping to a value estimated as dry air (the boundaries of the 25th percentile went beyond the limits for the assessment of "moderately dry air").

In the subarctic zone, the longest period was observed in the years when the air humidity went beyond the optimal value; the shortest was registered in the

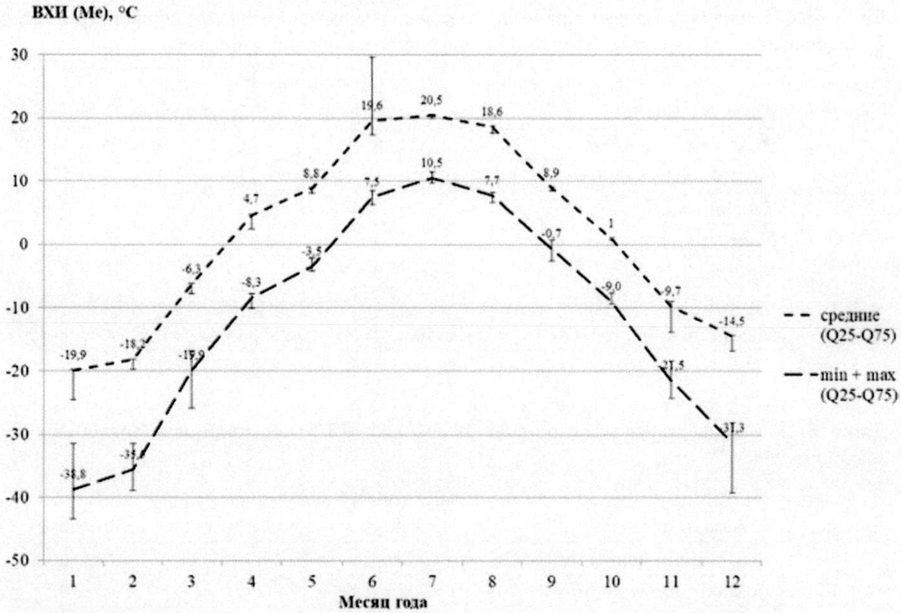

Fig. 5.26 Characteristics of the wind-chill index in Minusinsk conditions at different combinations of temperature and wind speed, °C

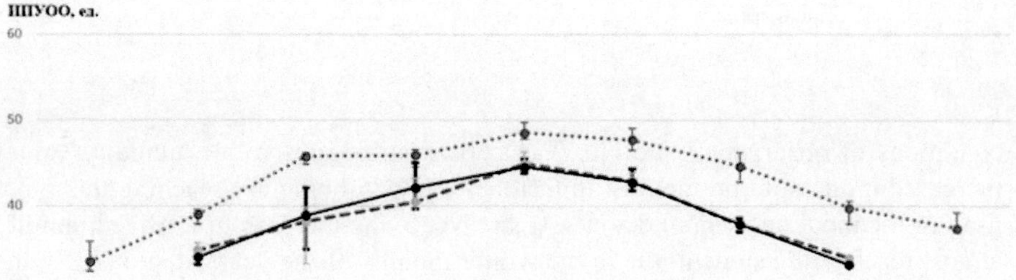

Fig. 5.27 IIBCC by the territory in the dynamics of annual observation characterizing the risk of cold exposure

Table 5.22 Indicators characterizing health risks when being outdoors in the compared territories, by median values and quartile intervals of average temperature and wind speed

No. item	Assessed criteria	Observation sites		
		Norilsk	Krasnoyarsk	Minusinsk
1	Risk period, months	8	6	5
2	Possible risk, months	1	1	2
3	Type of risk:			
	None	3	5	5
	Moderate	6	7	6
	Critical	3	–	1
4	Maximum values of IIBCC (Q75), units.	47.31–49.28	45.44–45.95	46.11–47.02

Table 5.23 Characteristics of wind-cold values and IIBCC at minimum temperature and maximum wind

No. item	Assessed criteria	Observation sites		
		Norilsk	Krasnoyarsk	Minusinsk
1	Minimum temperature, °C	−30.5 ± 1.6	−20.9 ± 1.5	−25.5 ± 1.9
2	Maximum wind, m/s	16.9 ± 1.0	7.9 ± 0.2	10.1 ± 0.5
3	Risk period, months	10	9	9
4	Type of risk:			
	None	2	3	3
	Moderate	3	6	6
	Critical	5 (4)	2 (3)	3
	Catastrophic	2 (3)	–	–

conditions of observation site No. 3. At observation site No. 1, humidity values increased in the autumn months, in October the Q75 boundary reaching the upper limit of the moderately humid zone. At site No. 3, the increase in relative humidity almost reached the same limits in the winter months of the year. At site No. 2, the increase in humidity stayed within the boundaries of moderately humid air.

When calculating the ET based on monthly averages of temperature and wind speed, it turned out that only in the conditions of the subarctic belt was the health risk determined for 2 months a year, in January and February. Me temperature values were −30.8 °C and −28.6 °C. The values of the quartile interval (Q25–Q75) were −30.8 °C to −28.7 °C and −29.2 °C to −28.2 °C, respectively. Under the conditions of site No. 2, the minimum value of temperature Me in January was 18.7 °C, Q25 to −26.0 °C, Q75 to −18.7 °C, site No. 3, respectively, −22.0 °C and −26.9 °C and −22.0 °C.

Calculation of ET by temperature and wind extremes showed the possibility of a more negative impact of physical factors on people in all territories of the region (Table 5.26). Thus, in the subarctic zone, the risk of frostbite was registered for 5 months a year, in site No 2 – 1 month, in site No 3 – 2 months a year.

Table 5.24 Characterization of the IIBCC by minimum temperature and maximum wind speed by observation area, units

No. item	Observation site	Months, M ± m											
		1	2	3	4	5	6	7	8	9	10	11	12
1	Norilsk	60.8 ± 1.2	59.4 ± 1.2	54.9 ± 1.5	49.8 ± 1.0	45.3 ± 0.6	39.8 ± 0.9	–	–	41.4 ± 1.1	47.4 ± 1.2	55.3 ± 0.9	56.2 ± 1.1
2	Krasnoyarsk	49.0 ± 1.1	49.0 ± 0.7		39.0 ± 0.4	37.2 ± 0.6	–	–	–	36.3 ± 0.5	39.5 ± 0.9	45.3 ± 0.9	47.1 ± 1.0
3	Minusinsk	50.7 ± 1.5	50.7 ± 0.8	45.3 ± 0.7	41.7 ± 0.4	40.1 ± 0.5	–	–	–	37.5 ± 0.5	41.6 ± 0.5	46.0 ± 0.8	49.1 ± 1.0

Table 5.25 Characteristics of air humidity according to relative humidity by object of observation, %

No. item	Assessed criteria	Observation site		
		№1	№2	№3
1	Moderately dry climate, number of months	2	4	5
	Minimum humidity, M ± m %	66.9 ± 3.2	57.5 ± 1.7	57.8 ± 1.2
	Months	VI–VI	III–VI	III–VII
	Me, %	71.5 (VII)	57.0 (IV)	58.0 (V)
	Q25, %	57.0	54.0	55.0
	Q75, %	74.0	58.0	60.0
2	Moderately humid, number of months	10	8	7
	Maximum humidity, M ± m%	81.7 ± 1.1	76.7 ± 1.1	80.2 ± 1.6
	Months of the year	X, XI	IX, XII	I–XII
	Me, %	72.5 (X)	76.5 (IX)	84.0 (XII)
	Q25, %	79.0	75.0	79.0
	Q75, %	85.0	80.0	84.0

Table 5.26 Characteristics of effective temperatures in the monitored territories by Me, °C (Q25-Q75)

No. item	Observation site	Month, M ± m					
		I	II	III	IV–X	XI	XII
1	Norilsk	−46.1 (−52.2 to −46.1)	−41.5 (−44.0 to −41.5)	−32.4 (−34.4 to −32.4)	No risk	−34.2 (−38.5 to −34.2)	−36.3 (−38.7 to −36.3)
2	Krasnoyarsk	−27.5 (−32.5 to −26.7)	No risk				
3	Minusinsk	−32.2 (−37.7 to −28.1)	−29.0 (−31.1 to −28.6)	No risk			

For ET in a temperate continental climate, the health risk criteria were less pronounced, with a longer period defined as "moderately cool" to "comfortable" (Table 5.27).

Table 5.27 Characterization of health risk by equivalent-effect temperature in the observation periods by climate zone, months

No. item	Type of risk	Subarctic	Continental
1	Threat of frostbite	5	1/2[a]
2	Very cold	1/2[a]	2/1[a]
3	Cold	1/2[a]	2
4	Moderately cold	1	–
5	Very cool	1	2
6	Moderately cool	–	2
7	Cool	2	1
8	Comfort (moderately warm)	1	2

[a]Borderline values

Chapter 6
Non-specific Prevention of Pre-disease States and Diseases When Adapting to a Maritime Climate

Our findings suggest that individuals with normal and poor nutritional status adapt differently to a maritime climate. This difference is particularly dramatic in the Black Sea area. Thus, there is a need to influence the metabolic processes by optimizing the diet of young males relocating to coastal regions. Formulations for multi-component foods containing vegetables, cereals, berries, and fruits were devised to this end. The products were obtained using modern technology, namely, cryogenic processing [86].

Cryogenic technology consisted of pretreatment, drying, and milling. Drying makes it possible to increase the concentration of biologically active substances (BAS) 8.0–12.0-fold by moisture removal (raw materials contain up to 90.0% of water before drying and 5.0–10.0% after it). Milling prevents raw material oxidation and releases biologically active elements bound to protein molecules. This way, elements are fully absorbed by the human body. Conventional milling cannot produce such an effect since it may cause raw materials to reach a temperature of over 200° and lose valuable BAS.

Freezing makes it possible to cut raw materials that cannot be handled otherwise because of changes in mechanical properties (embrittlement). This method also helps obtain powders with a particle size that cannot be ensured at higher temperatures (fine and extra-fine powdering). Another advantage is the preservation of biologically active elements in the ground product (the ingredients are not heated, and they are in an inert atmosphere). Finally, freezing prevents particle aggregation.

Cryogenic technology increases the bioavailability of BAS to the body, and it is a green method of production that does not use chemical processing.

According to the literature, cryogenic technology increases the content of carotenoids in *Sorbaronia mitschurinii* 18.4 times; of thiamine, 16.0 times; of vitamin E, 10.0 times. In cranberry, the content of thiamine rises 11.0-fold; of riboflavin, 12.0-fold.

In foods treated with cryogenic technology, the heat effect of enzymatic hydrolysis (enthalpy) rose in the presence of plant amylase by 13.8% for apples and by

A. V. Tarasov, R. S. Rakhmanov, *Marine Climate of Russian Coastal Territories*,
https://doi.org/10.1007/978-3-031-30951-9_6

Table 6.1 Vitamin and mineral content in the plant-based and plant-based protein foods, mg/100 g

BAS[a]	Plant-based food	Plant-based protein food
Vitamins:		
A	7.3	2.4
E	6.1	6.3
B_2	0.17	1.15
Mineral substances:		
Iron	13.2	3.8
Copper	0.3	0.13
Zinc	1.8	2.7
Manganese	1.4	0.3
Chromium	0.06	0.08

[a]identified by laboratory testing

19.5% for carrots. Nutrients were readily available this way and easily absorbed by the human body.

In addition, these foods have sorption properties with respect to heavy metals. For example, sorption of lead reached 52% in Jerusalem artichoke, 86% in pumpkin, 85% in blueberry, and 94% in parsley. For cadmium, these percentages were 67, 80, 71, and 86, respectively.

Formulations were developed for a plant-based protein natural concentrated food and its plant-based counterpart, both having a high content of antioxidant vitamins (A, E, C) and minerals (Table 6.1). Such foods have been reported to contain secondary metabolites of plants: flavonoids, plant polyphenols, carotenoids, low-molecular-weight nitrogen compounds, and organic acids [27]. The ingredients in the plant-based product were beetroot, rosehip, carrot, and apple, while the plant-based protein product consisted of non-fat cottage cheese, wheat germ, beetroot, and eggshells.

Subjects received the foods once daily for 15 days: 35 g of the plant-based protein food in powder form and 1.8 g of the plant-based food in tablets, 0.3 g each. Subjects with normal nutritional status were given only the plant-based food. They received the supplement in two courses 30 days apart. Individuals with poor nutritional status were given both plant-based and plant-based protein food in a single course.

The levels of cytokines (IL-1b and TNF-α) and immunoglobulins (IgG and A) in saliva were measured at the beginning of the study, along with the accumulation of polymorphonuclear leukocytes (PNL) of myeloperoxidase. At that stage, group members were exchanging microorganisms, including infectious agents. The obtained results showed that the young males had inflammatory reactions. In the control group, inflammations were observed throughout the study, that is, during 46 days.

Natural resistance mechanisms were activated in subjects with normal body weight. For example, the lysozyme level in saliva increased by 16.2% ($p < 0.05$).

The content of proinflammatory cytokine IL-1b did not alter and remained moderate. The tumor necrosis factor (TNF-α) was 97.2 ± 12.3 ng/mL: 21% increase by day 16 was followed by a return to the initial values by day 46. At the beginning of the study, the myeloperoxidase activity of neutrophils was 23.5 ± 3.4 ng/ml; by day 16, it rose by 46.7%; by day 46, it declined to 12.0 ± 3.1 ng/mL. Inflammatory reactions in this group, if registered, were moderate and dwindled toward the end of the study.

Resistance was also activated in the underweight group: the lysozyme concentration grew by 14.6% ($p < 0.05$). Lysozyme levels did not differ statistically in the treatment and control groups at the beginning of the study, and they remained unchanged until the end. Upon arrival in the region, subjects had an IL-1b concentration of 47.1 ± 5.4 ng/mL; 2 weeks later, it rose to 105.8 ± 10.8 ng/mL and decreased afterward. TNF-α showed a similar trend: it grew by 35.1% by day 16 and returned to the initial values by day 46. The initial level of myeloperoxidase activity of neutrophils was 29.3 ± 4.3 ng/mL; by day 16, it rose to 50.5 ± 14.5 ng/mL; by day 46, it returned to the initial value. The signs of pre-immune inflammation were more distinct in the underweight group than in subjects with normal nutritional status. The intake of functional foods was as effective against the signs of inflammation in the former as it was in the latter.

Significant within-group changes in salivary IgG immunoglobulin content occurred in the treatment group. In the control group, the IgG content increased ($p < 0.01$) at the time of the second and third tests, which pointed to the activation of the humoral immune system in response to a pathogen circulating as a result of contagion (the "sharing factor" was at play in the group). In the treatment and comparison groups, this measure changed slightly, that is, the factors of natural resistance of the body were activated. There were no statistically significant changes in immunocompetent cells.

The intake of functional foods increased functional body mass, normalized the functions of organs (respiration and the cardiovascular system), and raised vitamin levels. All this indicated metabolic stimulation in young males.

The body weight of subjects with normal nutritional status grew. The highest increase occurred 30 days after the course, reaching 3.03 ± 0.4 kg ($p < 0.05$).

The skinfold thickness also grew. Initially, it was 0.5 ± 1 cm; by the end of the plant-based functional food intake, 0.8 ± 0.2 cm; 30 days later (day 46), 0.9 ± 0.2 cm, $p < 0.01$.

The response to graduated exercise evidenced a significant rise in the adaptive capacity of the body. The proportion of individuals with an unsatisfactory response decreased from 33.3 to 8.3%. In the control group, the proportion of such subjects was 21.7–30.4%.

C-vitamin levels were satisfactory at all stages of the study. In the control group, 12.5% had vitamin saturation below normal (0.2 mg/h, the norm ranges from 0.3 to 0.7 mg/h). At the same time, the increase in urinary excretion of ascorbic acid measured in mg per hour reached 23.8–47.6% in the treatment group.

In the treatment group, the bactericidal activity of the skin increased from low values at the beginning of the study to the physiological norm at its end.

Morphofunctional improvements were observed in subjects with poor nutritional status receiving the plant-based protein food.

Respiratory disease incidence among young males with normal nutritional status on a diet including the functional food was 2.3 times lower than in the control group; the incidence of CAP was 2.7 lower; of URI, 2.9 times. In underweight subjects, the incidence of URI was 1.35 times lower than in the control group; of CAP, 1.7 times.

At the early stage of acclimatization, there was a correlation between the incidence of respiratory diseases and the start date of functional food intake. If the supplement was introduced in the first week, no cases of diseases were registered, while the incidence in the control group was 77.6 per 1.000 population. If the intake started in the second week, the incidence was 2.5 times lower than in controls. The introduction in the third week led to a 15.3% difference.

The plant-based functional food was employed to prevent digestive diseases. Two series of studies were carried out. Subjects were prescribed beetroot and celery or sprouted wheat and oats. In some cases, the intake of a liquid eubiotic (concentrated bifidumbacterin) was authorized. In one series, subjects received half-doses of the plant-based functional food and liquid concentrated bifidumbacterin.

Gastroduodenoscopy and histological and microscopic studies of samples from subjects in experimental groups were conducted. They did not show any significant increase in the cases of gastric mucosa inflammation after five or five and a half months, compared to the early stage of acclimatization. Gastroduodenoscopy confirmed a rise by 5%; histology, by 10%. The frequency of *H. pylori* infection increased by 5%. In controls, the rise in the detectable signs of gastritis, gastritis, and bulbitis was 30 and 20%. The frequency of obtaining histological evidence for inflammation increased by 30%; of detecting *H. pylori* in smears by 20%. In the first group, 20–50 bacteria per HPF were detected; in the second, over 50. The rate of occurrence of signs of superficial gastritis went up in the treatment group. In the control group, superficial gastritis was detected 1.5 times more often than at the beginning of the study; atrophic gastritis, 1.3 times. Some subjects exhibited signs of granular gastritis.

The intake of probiotics as prophylaxis produced a positive effect absent in the control group. By day 91, visual signs of the inflammation of gastric mucosa were registered as often as initially in the group receiving bifidumbacterin. However, the rate of occurrence of gastric and duodenal mucosa inflammation signs dropped. The frequency of histological evidence for inflammation in samples decreased. Although the prevalence of *H. pylori* grew by 5%, there were fewer microorganisms per HPF. Histology proved a decrease in the number of subjects with signs of atrophic gastritis.

Using a complex method for preventing acclimatization disorders and Helicobacteriosis at the early stages of adapting to a new environment resulted in weight loss in controls by day 41 and weight gain by 1.5 kg in the treatment group ($p > 0.05$). The RUAC did not change in the control group but increased in the treatment group. The dominant hand strength did not change in the control group and significantly increased, by 3.7 kg, in the treatment group by day 41. The vital capacity and the breath-hold test results did not change in the control group and went up by 22.2% ($p < 0.01$) and 30.9%, respectively, in the treatment group ($p < 0.01$). The control group showed a tendency toward reduced cardiovascular function both at rest and after graduated exercise. In the treatment group, the function improved.

C-vitamin levels did not alter in the control group but increased 1.4-fold in the treatment group by day 11 and remained unchanged for a month.

Gastroduodenoscopy revealed a 30% increase in the number of subjects with signs of inflammation of the gastric mucosa and duodenum.

Histological evidence for inflammation was 40% more widespread. The number of subjects with signs of superficial gastritis increased 1.5-fold; of atrophic, three-fold. One person had granular gastritis signs. The frequency of microscopic confirmation of *H. pylori* grew by a factor of 1.8. The number of bacteria in samples increased as well.

There was no increase in antral mucosa inflammation signs in the treatment group, while the proportion of subjects with inflammation signs in the gastric and duodenal mucosa decreased. There was no increase in histological evidence for inflammation in the gastric mucosa. The number of persons exhibiting signs of atrophic gastritis decreased, while that of subjects with symptoms of other types of gastritis did not alter. The 10% rise in the number of people diagnosed with *H. pylori* was not accompanied by a growing number of bacteria per HPF.

The effect of the Revit multivitamin supplement, the plant-based functional food, and bifidumbacterin was estimated in individuals with normal nutritional status in the maritime climate of the Caspian Sea [87].

The natural resistance of the body was evaluated based on the antimicrobial resistance of the skin. Initially, resistance in the group receiving the Revit supplement was statistically higher than in the group given the plant-based functional food alone. The time of skin exposure to test microbes was 3 min. The values in controls and the Revit and bifidumbacterin groups did not differ significantly.

There were no statistically significant differences in the bactericidal indices at the end of the treatment when the time of exposure increased to 5 and 10 min. However, 15-min exposure confirmed a statistically significant growth in bactericidal indices in the treatment group, compared to controls. The indices were 7.8% higher in the Revit group ($p < 0{,}05$); 6.5% in the bifidumbacterin group ($p < 0{,}05$); 12.8% in the functional food group ($p < 0{,}01$). Compared to the initial values, antimicrobial resistance of the skin increased in the Revit group by 6.4% ($p < 0.05$); in the bifidumbacterin group by 11.6% ($p < 0.01$); in the functional food group by 14.9% ($p < 0.01$). In the latter, the increase in bactericidal activity was statistically higher than in the bifidumbacterin group, by 6.3%.

The intake of the multivitamin supplement, bifidumbacterin, and the functional food helped normalize microflora in the large intestine. The most impressive results were achieved when subjects received bifidumbacterin. Against the background of a slight decrease in the content of *E. coli* with normal enzymatic activity, that of *E. coli* with the reduced enzymatic activity of lactose-negative *E. coli* diminished. Proteus and Staphylococcus aureus were eliminated, and the content of enterococci dropped by a quarter. The content of *Bifidobacteria* increased, along with that of lactobacilli. The plant-based functional food showed the second-best result.

Table 6.2 Indicators of parapharmaceutical effectiveness in preventing skin and subcutaneous tissue diseases (SSTD)

Group	Incidence	Subjects diagnosed with the conditions by the end of treatment, %	Reduction compared to initial values, %
Control	Total	20.0	16.0
	SSTD[a]	12.0	–
Revit	Total	8.0	32.0
	SSTD	12.0	12.0
Plant-based functional food	Total	28.0	16.0
	SSTD	16.0	12.0
Bifidumbacterin	Total	16.0	16.0
	SSTD	4.0	8.0

[a]skin and subcutaneous tissue diseases

The incidence of skin and subcutaneous tissue diseases in subjects was measured at the early stage of acclimatization. Three months before the study, 28–40% of individuals in the treatment groups and 36% in the control group were diagnosed with the diseases. After the intake of nutraceuticals and parapharmaceuticals and 3 months later, the incidence in the groups receiving bifidumbacterin, Revit, or the functional food was 1.4, 1.6, and 1.8 times lower than in controls (Table 6.2). The best result was in the functional food group.

A comparative assessment of treatment methods showed that option 1 (standard) was more effective than the application of topical medications coupled with Revit (option 2) or topical medications combined with nutraceuticals and the parapharmaceutical bifidumbacterin (option 3). In the 5 months following the study, only 4% of subjects relapsed. This proportion was three times lower than when using the standard option and four times when combining topical medication with multivitamin supplements. The best results were obtained when the plant-based function food was included in the standard treatment: clinical improvement took place earlier, treatment periods were shorter, and no relapses were observed (Table 6.3).

A plant-based functional food consisting of watermelon seeds, rose hips, oats, kelp, and egg whites was included in the diet of subjects to ease their adaptation to the Baltic climate and prevent pre-disease states. Three tablets (0.3 g each) were taken twice a day at breakfast and supper for 15 days. An equal group of young males was chosen as control [87].

The plant-based protein functional food had a considerable content of vitamins C, E, K, B_2, and A, as well as iron, zinc, manganese, copper, and chrome (Table 6.4).

At the beginning of the study, subjects in both treatment and control groups had vitamin A levels above the reference limits. No statistically significant changes took place in controls (p 1:2 = 0.265), yet a 21.3% ($p = 0.002$) increase occurred by the end of the functional food intake in the treatment group. By the end of the study, the levels rose by 23.6% ($p = 0.000$).

Table 6.3 Comparative effectiveness of different ways to treat skin and subcutaneous tissue infections

Measure	Comparison groups			
	1	2	3	4
Clinical improvement, day	6–7	10–12	8–9	4–5
Sickness absence, days	8.6 ± 0.5	16 ± 0.5	12.2 ± 0.6	5.7 ± 0.6
Relapses, %	12.0	16.0	4.0	0

Table 6.4 Contents of biologically active substances in the functional food, mg/100 g

Minerals				
Copper	Zinc	Iron	Manganese	Chrome
0.58	9.86	71.0	2.85	0.16
Vitamins				
A	E	B_2	K	C
0.027	3.69	0.3	0.51	89.91

Initially, subjects in experimental groups had very similar vitamin E levels. This rate declined by 12.5% ($p = 0.019$) in the control group at the end of the functional food intake and did not change onward ($p = 0.049$). Vitamin E deficiency was observed in 40% of subjects. This proportion rose to 90% ($\varphi = 3.567$) after 15 days and remained unchanged onward (80%, $\varphi = 2.672$). In experimental groups, this rate went up by 10.6% ($p = 0.196$) and 16.9% ($p = 0.0.17$) toward the end of the study. Vitamin E levels rose in 65 and 75% of subjects, as evidenced by the second and third tests.

In both groups, vitamin B1 levels went down slightly by the end of the study. In the control group, they were 25.9 ($p = 0.002$) and 17.2% ($p = 0.039$) lower than in the treatment group.

Vitamin B2 levels did not change throughout the study in controls. Yet, they increased by 13.5% in the treatment group ($p = 0.044$). At the end of the plant-based protein food intake, this rate was statistically higher than in the control group, by 17.9% ($p = 0.042$). At the end of the study, this difference increased to 28.8% ($p = 0.001$).

Toward the end of the functional food intake, the proportion of controls with low copper levels grew ($\varphi = 4.234$). In 65.5% of subjects in the treatment group, copper levels either rose or remained unchanged within the normal range (differences between treatment and control groups were statistically significant, $\varphi = 3.893$). Overall, by the end of the functional food intake, copper levels in the treatment group were statistically higher than in the control group, $p = 0.046$.

By the end of the study, a decrease in zinc levels, albeit within reference limits, occurred in 50% of controls ($\chi 2 = 10.8$, $p = 0.001$) and 20% of subjects in the treatment group. Zinc levels rose in 75% of the treatment group ($\varphi = 3.004$, the difference was statistically significant).

Iron levels did not differ initially in the experimental groups. Nor did statistically significant changes occur in the control group over time: p1-r2 = 0.078; p1-r3 = 0.106.

In the treatment group, 1 month after the functional food intake, iron levels were statistically higher than in the initial state (by 23.5%, $p = 0.023$).

No changes in calcium levels took place in any of the groups.

There was a statistically significant reduction in magnesium levels in the control group: in the second test, it was 22.6% below the initial level ($p = 0.000$); in the third, by 13.6% ($p = 0.000$). A decrease was observed in 100 and 90% of subjects, respectively. In the treatment group, however, magnesium levels rose. Although they were statistically lower than in the control group initially, there was no difference at the end of the study. By the end of the functional food intake, magnesium levels went up by 8% ($p = 0.004$); by the end of the study, by 36.3% ($p = 0.000$). In the second test, an increase in magnesium levels was detected in 60% of subjects, while it remained unchanged in 30%. At the end of the study, a rise occurred in all participants.

Serum phosphorus levels were within the reference limits in the control group. Nevertheless, a decrease occurred afterward in 66.7% of the subjects ($\chi2 = 14.405$, $p = 0.0001$). At the end of the study, phosphorus levels below the initial values were observed in 33.3% of controls ($\chi2 = 14.405, p = 0.0001$). In the treatment group, this rate rose by 16.1% ($p = 0.000$) by the end of the functional food intake and by 21.3% ($p = 0.000$) when the study was completed.

Serum potassium, sodium, and chloride levels were within the reference limits throughout the study. No statistically significant changes occurred at any stage. Yet, there was a tendency toward a decrease in chloride levels in the control group.

An increase in total protein levels in the blood was detected in 60% of the treatment group by the end of the functional food intake; in another 5%, they remained unchanged. In the control group, protein levels decreased in 63.6% of subjects ($\chi2 = 14.737, p = 0.000$). By the end of the study, 40% of subjects had protein levels below the initial values ($\chi2 = 7.265, p = 0.007$).

There was no significant change in the peroxide level, an indicator of the oxidative stress state, in the control group by the end of the functional food intake. Yet serum antioxidant activity decreased by 11.2% ($p = 0.024$), pointing to a worsening antioxidant status. In the treatment group, peroxide levels fell over time: by 33.7% by the end of the functional food intake ($p = 0.044$) and by 62.8% 1 month after the course ($p = 0.000$). Serum antioxidant activity, on the contrary, increased by 24.9% ($p = 0.000$) by the end of the study. All this testified to an attenuation of oxidative stress signs.

In the control group, the proportion of individuals with a low or average degree of oxidative stress decreased, while it rose in the treatment group.

Serum IgG levels were within the reference limits in both groups. A 27.8% decrease in this measure occurred in the control group by the end of month 1. A statistically significant increase, by 43.6% ($p = 0.014$), was observed in the treatment group. The proportion of subjects with normal IgG levels decreased twofold in the control group by the end of the study, while it remained unchanged in the treatment group.

Serum IgA levels decreased statistically, by 37.5% ($p = 0.000$), in the control group by the end of the study but did not change in the treatment group (p1-r2 = 0.542; p2-r3 = 0.271). The proportion of persons with normal IgA levels declined in the control group, while the percentage of individuals with low IgA levels increased. In the treatment group, the proportion of subjects with normal IgA levels grew, and the share of those with low levels declined.

At the beginning of the study, C-reactive protein levels were within the normal range in both groups, and no statistically significant changes occurred in controls (p1-r2 = 0.618; p1-r3 = 0.825). In the treatment group, a decrease in serum protein levels took place. Although it was not statistically significant by the end of the functional food intake ($p = 0.698$), it became so toward the end of the study when the decline reached 67.2% ($p = 0.005$). A considerable proportion of subjects with C-reactive protein levels above the normal limits was in the treatment group in the first test and controls at the end of the study.

The testosterone level in the control group decreased over time: by 29.6% ($p = 0.000$) by the end of the functional food intake and by 32.8% ($p = 0.000$) 1 month later. The hormone level in the treatment group fell only at the end of the study, by 14.3%. This decrease, however, was not statistically significant ($p = 0.19$). The cortisol level was 60% above the reference limits in the control group and by 25% at the end of the functional food intake.

There was a statistically significant increase, by 6.2%, in the weight of treatment group subjects by the end of the observation period. This rise occurred in all participants. In the control group, 20% of subjects lost weight by the end of the study. The dominant hand strength did not change in that group, while it increased in the experimental group (by 7.3% compared to the initial value).

Chapter 7
Conclusion

The maritime climate of Russian regions has a health-promoting effect, widely used to improve the health of children and adults. Yet natural and climatic factors can also have a negative influence. Data supporting these two theses were obtained in this study. The data on the morphofunctional states of young males at early stages of job adaptation in a maritime climate suggest an essentially negative influence of the environment.

We drew on recent knowledge of acclimatization to emphasize the role of the initial nutritional status in preserving homeostasis and describe the contribution of vitamin and mineral levels. Pronounced shifts in the large intestine microbiota composition, the state of gastric and duodenal mucous, and the presence of *H. pylori* bacteria in underweight subjects were linked to the bactericidal activity of the skin.

Acclimatization affected metabolic processes in the body at the early stages of acclimatization. This change manifested in weight loss, waning hand muscle strength, and falling vitamin, mineral, total protein, protein fraction, and γ-globulin levels. Acclimatization was shown to reduce immunoglobulin production, affect the natural resistance of the body, and cause pre-immune inflammation signs. There was also a rise in oxidative stress phenomena. Peroxide levels in serum measured in the first days spent in the Baltic indicated high oxidative stress and an average level of serum antioxidant activity. Serum antioxidant activity continued to decrease by the end of the study, pointing to a worsening antioxidant status. The proportion of individuals with low and medium oxidative stress levels fell, while the percentage of those with high levels rose. The antioxidant activity of blood serum was decreasing: the share of persons with low and average values was growing, and that of individuals with a high activity decreasing.

The risk of diseases associated with malfunctions of natural resistance mechanisms – URIs, CAP, furunculosis, abscesses, phlegmon, pyoderma, and gastric and duodenal ulcers – grew. The findings by Bogomolov corroborate our results: young males changing their place of residence are especially susceptible to CAP. Those

A. V. Tarasov, R. S. Rakhmanov, *Marine Climate of Russian Coastal Territories*,
https://doi.org/10.1007/978-3-031-30951-9_7

relocating in the autumn had difficulty adapting to the autumn-winter outbreak of CAP. They had a more severe course of disease.

In people strange to a maritime climate, acclimatization disorders and overall incidence were significantly higher than in those living in the weather and climatic conditions for a long time. According to Bogomolov, CAP in young males relocating to the Primorsky region is associated with severe clinical symptoms, including intoxication and respiratory tract lesion, a considerable number of complications, and a longer duration of treatment.

In Kamchatka, the most prevalent diseases among 25–50-year-old males working together were respiratory disorders. The other most common conditions were diseases of the musculoskeletal system and connective tissue, disorders of the eye and adnexa, infectious and parasitic diseases, diseases of the digestive system, injuries and other health conditions due to an external cause, and diseases of the urogenital system. Together, they accounted for 77.1% of the total morbidity.

In Chukotka, the most prevalent category was respiratory diseases, while digestive diseases ranked second. The seven most common categories comprised 92% of the total morbidity. These were diseases of the musculoskeletal system and connective tissue, circulatory system diseases, injuries and other health conditions due to an external cause, diseases of the urogenital system, and skin and subcutaneous tissue diseases.

In Magadan, the two categories that ranked the highest (digestive and respiratory diseases) did not differ in incidence ($p = 0.423$). Overall, six categories (digestive diseases, respiratory diseases, diseases of the musculoskeletal system and connective tissue, diseases of the eye and adnexa, diseases of the endocrine system, eating disorders and metabolic disorders, and diseases of the urogenital system) accounted for 85.8% of the total morbidity.

Morbidity in Magadan was 2.4 times higher than in Kamchatka and 2.9 times than in Chukotka.

These data suggest that the diet of young males adapting to a new climate should be adjusted to include foods having an antioxidant effect, correcting protein and energy deficiency, and activating natural resistance mechanisms, including colonization resistance.

The role of nutrition as a factor in prophylaxis and acclimatization to extreme conditions has been emphasized by Pokrovsky. He writes that each of many hundreds and thousands of substances found in food has a degree of biological activity. It is thus essential to view food not only as a source of energy and nutrients but also as a pharmacological complex [88].

Formulations for plant-based protein foods were devised. The ingredients were processed by low-temperature drying technology to increase the concentrations of biologically active substances. The concentrated foods had high concentrations of antioxidant vitamins (A, E, C) and minerals. Concentrated liquid bifidumbacterin was used to restore the normal microbiocenosis.

An increase in vitamin and mineral levels and additional intake of animal protein by underweight subjects led to the normalization of metabolic processes, which was confirmed by hormone measurements. In subjects taking concentrated products, the levels of the catabolic hormone cortisol decreased by the end of month 3, while those of the anabolic hormone testosterone did not change significantly. In the control group, testosterone levels fell by 32.8% by the end of month 3. Cortisol levels were above the reference limits in 60% of controls by the end of month 1, compared to 25% of subjects receiving the functional food. The supplement had a positive effect on the structures, functions, adaptive capacity, and natural resistance. The antioxidant status increased, while *H. pylori* bacteria was detected less often. Functional food intake helped prevent gastric and duodenal inflammation and normalize large intestine microbiota.

The incidence of CAP was 2.7 times lower among subjects with normal weight; of URIs, 2.9 times. In underweight subjects, the incidence of URIs was 1.35 times lower; of CAP, 1.7. The intake of products with a high content of biologically active substances by patients with skin and subcutaneous tissue infections shortened the duration of treatment and prevented relapses.

Natural products containing vitamins, minerals, and secondary metabolites of plants had a more pronounced preventive effect than vitamin fortification and the intake of bifidumbacterin. The bactericidal activity of the skin increased, the number of subjects diagnosed with a condition on a first-time visit diminished, and the incidence of skin and subcutaneous tissue infections declined. Moreover, the functional food had a longer after-effect.

At the early stage of acclimatization, there was a correlation between the incidence of respiratory diseases and the start date of functional food intake. If the supplement was introduced in the first week, no cases were registered, while the incidence in the control group was 77.6 per 1.000 population. If the intake started in the second week, the incidence was 2.5 times lower than in controls. The introduction in the third week translated into a 15.3% difference.

Such positive effects were due to optimizing the diet of young males adapting to the maritime climate of the Black, Caspian, and Baltic Sea regions.

The weather and climate of Russia's coastal regions negatively affect the morphofunctional state of people going through job adaptation and cause morbidity to increase. This fact should be taken into account by management wishing to promote the health and fitness of employees and medical professionals seeking to prevent acclimatization disorders and related diseases.

References

1. Kaznacheev, V.P. Sovremennye aspekty adaptacii/V.P. Kaznacheev. M.: Nauka, 1980. 192 s.
2. Koshheev, V.S. Fiziologija i gigiena individual'noj zashhity cheloveka ot holoda/V.S. Koshheev. M.: Medicina, 1981. 288 s.
3. Novozhilov, G.N. Gigienicheskaja ocenka mikroklimata/G.N. Novozhilov. L.: Medicina, LO, 1987. 112 s.
4. Azhaev, A.N. Fiziologo-gigienicheskie aspekty dejstvija nizkih temperatur na organizm cheloveka/A.N. Azhaev, I.A. Berzin, S.A. Deeva. M.: Medicina, 2008. 120 s.
5. Chvyrev, V.G. Teplovoj stress/V.G. Chvyrev, A.N. Azhaev, G.N. Novozhilov. M.: Medicina, 2000. 296 s.
6. Adol'f, Je. Fiziologija cheloveka v pustyne/Je. Adol'f. M.: Mir, 1952. 360 s.
7. Galanin, N.F. Luchistaja jenergija i ejo gigienicheskoe znachenie/N.F. Galanin. L.: Medicina, 1969. 184 s.
8. Malysheva, A.E. Gigienicheskie voprosy radiacionnogo teploobmena cheloveka s okruzhajushhej sredoj/A.E. Malysheva. M.: Medgiz, 1963. 244 s
9. Sapov, I.A. Nespecificheskie mehanizmy adaptacii cheloveka/I.A. Sapov, V.S. Novikov. L.: Nauka, 1984. 148 s
10. Vjazickij, P.O. Adaptacija molodogo voina k uslovijam voennoj sluzhby i profilaktika dizadaptacionnyh rasstrojstv/P.O. Vjazickij [i dr.]. M.: Voenizdat, 1990. 28 s.
11. Rahmanov, R.S. Ocenka vitaminno-mineral'nogo statusa i gumoral'nogo immuniteta junoshej pri psihosocial'noj adaptacii i akklimatizacii/R.S. Rahmanov, S.A. Razgulin, A.V. Tarasov// Medicinskij al'manah. 2014. № 2(32). S. 149–151.
12. Ivanov, V.P. Medicinskaja jekologija/V.P. Ivanov, N.V. Ivanova, A.V. Polonikov. SPb.: SpecLit, 2012. 320 s.
13. Ionov, P.K. osnovy detskoj kurortologii. Krasnodar: 'Sov. Kuban', 2000. 336 s.
14. Ujanaeva A.I., Vlijanie klimata i pogody na mehanizmy formirovanija povyshennoj meteochuvstvitel'nosti/A.I. Ujanaeva, Ju.Ju. Tupicyna, M.A. Rassulova//Voprosy kurortologii, fizioterapii i lechebnoj fizicheskoj kul'tury. 2016;93:5:52–57.
15. Grigor'eva E.A. Klimaticheskie uslovija Dal'nego Vostoka kak faktor razvitija boleznej organov dyhanija. Regional'nye problemy. 2017; 20 (4):79–8516.
16. Veremchuk, L.V., Yankova V.I., Vitkina T.I., Nazarenko A.V., Golokhvast K.S. Urban air pollution, climate and its impact on asthma morbidity. Asian Pacific Journal of Tropical Biomedicine. 2016; 6 (1): 76–79.
17. Dubrovskaja S.V. Meteochuvstvitel 'nost' i zdorov'e. M.: RIPOL klassik, 2011. 180 s

A. V. Tarasov, R. S. Rakhmanov, *Marine Climate of Russian Coastal Territories*,
https://doi.org/10.1007/978-3-031-30951-9

18. Tarasov, A.V. The role of plant-based protein functional food in preventing acute respiratory disease: A case study/A.V. Tarasov, L.A. Perminova, Z.L. Malakhova [et al.]//Nutrients. 2021. Vol. 13. No 6. P. 2116. https://doi.org/10.3390/nu13062116.
19. Vorob'eva, L.V.Gigiena, sanologija, jekologija: uchebnoe posobie/L.V. Vorob'eva; pod red. L. V. Vorob'evoj. SPb.: SpecLit, 2011. 255 s.
20. Matjuhin, V.A. Bioklimatologija v uslovijah mussonov/V.A. Matjuhin. L., 1971. 138 s.
21. Beljuchenko, I.S. Jekologija Krasnodarskogo kraja (Regional'naja jekologija): uchebnoe posobie/I.S.Beljuchenko. Krasnodar: Kubanskij GAU, 2010. 356 s.
22. Bardov, V.G. Gigiena klimata: ucheb. posob. po obshhej gigiene/ V.G. Bardov. Kiev, 1990. 136 s.
23. Rumjanceva, G.I. Obshhaja gigiena/G.I. Rumjanceva, M.P. Voroncova. M.: Medicina, 1990. 288 s.
24. Zhivoderov, V.M. Osobennosti techenija infarkta miokarda v uslovijah Primor'ja/ V.M. Zhivoderov. M., 1972. 29 s.
25. Rubin, A.D. Nauchno-organizacionnye osnovy optimizacii profilakticheskoj raboty s det'mi v uslovijah evropejskogo Severa Rossii: avtoref. dis. … dokt. med. nauk/A.D. Rubin. SPb., 2009. 29 s.
26. Arhangel'skij, V.I. Gigiena s osnovami jekologii cheloveka: uchebnik/V.I. Arhangel'skij, T.A. Kozlova [i dr.]; pod red. P.I. Mel'nichenko. M.: GJeOTAR-Media, 2010. 752 s.
27. Maljarenko T.N. Klimat i zdorov'e cheloveka. Soobshhenie 2/T. N. Maljarenko, A. T. Bykov, Ju. E. Maljarenko, A. A. Shhetkin//Valeologija. 2008. № 4. S. 4-11.28.
28. WHO. Quantitative risk assessment of the effects of climate change on selected causes of death, 2030s and 2050s. Geneva: World Health Organization. 2014.
29. Marchetti E., Capone P., Freda D. Climate change impact on microclimate of work environment related to occupational health and productivity. Ann Ist Super Sanita. 2016;52 (3):338–342.
30. Revich B.A. Shaposhnikov D.A., Anisimov O.A., Beloluckaja M.A. Vlijanie temperaturnyh voln na zdorov'e naselenija v gorodah Severo-Zapadnogo regiona Rossii. Problemy prognozirovanija. 2019;3: 127–134.
31. Shipko Ju.V., Shuvakin E.V., Ivanov A.V. Obobshhennyj bioklimaticheskij pokazatel' bezopasnosti rabot na otkrytom vozduhe v surovyh pogodnyh uslovijah. Vestnik KVGU, Serija: geografija. Geojekologija. 2015; 3:33–39.
32. World Health Report 2002. Reducing Risks, Promoting Healthy Life. World Health Organization, Geneva, 7–14.
33. Vinogradova, V. V. Universal'nyj indeks teplovogo komforta na territorii Rossii/V. V. Vinogradova//Izvestija Rossijskoj akademii nauk. Serija geograficheskaja. 2019. № 2. S. 3–19. https://doi.org/10.31857/S2587-5566201923-19.
34. Im J S, Han J, Lee M, et al. A study of skin characteristics according to humidity during sleep. Skin Res Technol. 2019;25(4):456–460.
35. Sunwoo Y, Chou C, Takeshita J, et al. Physiological and subjective responses to low relative humidity. J Physiol Anthropol. 2006;25 (1):7–14.
36. Hashiguchi N, Tochihara Y. Effects of low humidity and high air velocity in a heated room on physiological responses and thermal comfort after bathing: an experimental study. Int J Nurs Stud. 2009;46 (2):172–80.
37. Marchetti E., Capone P., Freda D. Climate change impact on microclimate of work environment related to occupational health and productivity. Ann Ist Super Sanita. 2016;52 (3):338–342.
38. Cariappa M P, Dutt Manohar, Reddy K P, et al 'Health, Environment and Training': Guidance on conduct of physical exertion in hot and humid climates. Med J Armed Forces India. 2018;74 (4):346–351.
39. Kjellstrom T. Crowe Jennifer Climate change, workplace heat exposure, and occupational health and productivity in Central America. Int J Occup Environ Health. 2011; 17(3):270–81.
40. Suntronwong N, Vichaiwattana P, Klinfueng S, et al. Climate factors influence seasonal influenza activity in Bangkok, Thailand. PLoS One. 2020;15 (9):e0239729.
41. Alenikova A.Je., Tepisova E.V. Analiz izmenenij gormonal'nogo profilja muzhchin g. Arhangel'ska v zavisimosti ot faktorov pogody. Vestnik Severnogo (Arkticheskogo) federal'nogo universiteta. Serija: Mediko-biologicheskie nauki. 2014;3:5–15.

42. Bocharov M.I. Termoreguljacija organizma pri holodovyh vozdejstvijah (obzor). Soobshhenie I. Vestnik Severnogo (Arkticheskogo) federal'nogo universiteta. Serija: Mediko-biologicheskie nauki. 2015;1:5–15.
43. Grigor'eva E. A., Hristoforova N. K. Bioklimat Dal'nego Vostoka Rossii i zdorov'e naselenija. Jekologija cheloveka. 2019;5:4–10.
44. Chashhin V.P., Gudkov A.B., Chashhin M. V, i dr. Prediktivnaja ocenka individual'noj vospriimchivosti organizma cheloveka k opasnomu vozdejstviju holoda. Jekologija cheloveka. 2017; 5:3–13.
45. Ivankov A. Explainer: What is Wind Chill? What are Its Effects? Posted on January 31, 2019. URL: https://www.profolus.com/topics/explainer-what-is-wind-chill-what-are-its-effects.
46. Gruza G. V. Nabljudaemye i ozhidaemye izmenenija klimata Rossii: temperatura vozduha. M.: IGKJe Rosgidrometa i RAN, 2012. 193 s.
47. Sinicyn I.S., Georgica I. M., Ivanova T. G. Bioklimaticheskaja harakteristika territorii v mediko-geograficheskih celjah. Jaroslavskij pedagogicheskij vestnik. 2013; 3 (4): 279–283.
48. Kuzjakina M.V., Gura D.A. Ocenka komfortnosti bioklimaticheskih uslovij Krasnodarskogo kraja s primeneniem GIS-tehnologij. Jug Rossii: jekologija, razvitie. 2020;15(3):66–76
49. Bahmudov, G.G. Gigienicheskoe obosnovanie optimizacii raciona pitanija pogranichnikov pri adaptacii k sluzhbe v kontrastnyh klimaticheskih uslovijah (Prikaspijskaja nizmennost' – srednegor'e): avtoref. dis. … kand. med. nauk/G.G. Bahmudov. N. Novgorod, 2005. 25 s.
50. Kandror, I.S. Ocherki po fiziologii i gigiene cheloveka na Krajnem Severe/I.S. Kandror. M.: Medicina, 1968. 280 s.
51. Zerenkov, P.A. Gigienicheskoe obosnovanie meroprijatij po sohraneniju i ukrepleniju zdorov'ja prizyvnoj molodezhi i voennosluzhashhih Tihookeanskogo flota: avtoref. dis. … kand. med. nauk/P.A. Zerenkov. Vladivostok, 2009. 19 s.
52. Surikova, L.E. Kliniko-biologicheskaja harakteristika stenokardii v uslovijah mussonnogo klimata g. Vladivostoka/L.E. Surikova. Vladivostok, 1971. 23 s.
53. Aubert, M. Impacts biologiquesdes aerosolsmarins/M. Aubert, J. Aubert//XV Congr. Intern. Thalassotherap. France, 1972, p. 1.
54. Fregschmidt, P. Endokrinologische Aspekte bei Meeres kunen/P. Fregschmidt//Z. angew. Bäder- u. Klimaheilk.,1970, Bd 17, N2, S. 135.
55. Pertot, V. Oplivob morskih dejiami kovnalaes kiorganizem in patofiziologi kadogajanja v niem/V. Pertot//Zdravstvenivestnik, 1971, v. 40, N4, s. 125.
56. Gabovich, R.D. Gigiena/R.D. Gabovich, S.S. Poznanskij, G.X. Shahbazjan. 3-e izd., pererab. i dop. Kiev: Vishha shkola, Golovnoe izd-vo, 1984. 320 s.
57. Vorob'ev, A.A. Osobennosti fiziologicheskih reakcij organizma morjakov pri rezkoj smene klimaticheskih rajonov/A.A. Vorob'ev. M., 1970. 27 s.
58. Bojko, E.L. Izuchenie akklimatizacionnyh fiziologicheskih sdvigov u morjakov pri bystroj smene klimaticheskih uslovij v processe dlitel'nogo plavanija na morskih sudah: avtoref. dis. … kand. med. nauk/E.L. Bojko. Odessa, 1972. 22 s.
59. Bondarev, G. Ja. Osobennosti pitanija v razlichnyh klimaticheskih regionah i v razlichnyh uslovijah truda/G.Ja. Bondarev, O.A. Shovkopljas//Rukovodstvo po gigiene vodnogo transporta. M., 1974. S. 238–240.
60. Ionov, P.K. Jeffektivnye principy i metody jendojekologicheskoj mediciny v lechenii i ozdorovlenii: opyt feder. kurorta Anapa/P.K. Ionov. Krasnodar: Sovet. Kuban', 2004. 319 s.
61. Boksha, V.G. Klimat lechit: spravochnik/V.G. Boksha, Ja.M. Bershickij. 3-e izd., dop. Simferopol': Tavrija, 1990. 55 s.
62. Lotyshev, I.P. Geografija Krasnodarskogo Kraja/I.P. Lotyshev. Krasnodar: Pechat. dvor Kubani, 1999. 135 s.
63. Shherbakov, D.B. Vzaimosochetanie prirodnyh lechebnyh mineral'nyh vod, peloideterapii, nutricevtikov i talassoprocedur v sovremennoj sisteme vosstanovitel'nogo lechenija na kurorte Anapa detej s razlichnymi formami atopicheskogo dermatita: avtoref. dis. …kand. med. nauk./ D.B. Shherbakov. Sochi, 2004. 17 s.

64. Voronin, Ja.M. Izmenenie gazoobmena, temperatury kozhi i korkovyh reakcij u bol'nyh i zdorovyh ljudej v period akklimatizacii k morskomu klimatu Jalty i Feodosii/Ja.M. Voronin [i dr.]//Problemy jeksperimental'noj kurortologii. M., 1968. S. 72–82.
65. Basalyga, V.N. Gigienicheskoe obosnovanie meroprijatij po nespecificheskoj profilaktike boleznej organov dyhanija v organizovannyh voinskih kollektivah: avtoref. dis. ... kand. med. nauk/V.N. Basalyga. N. Novgorod, 2003. 20 s.
66. Vishnevskaja, E.P. Obshhaja gigiena/E.P. Vishnevskaja [i dr.]. M.: Medicina, 1973. 328 s.
67. Shipicin, V.Ju. Jepidemiologo-gigienicheskie problemy helikobakterioza v voinskih kollektivah: avtoref. dis. ... kand. med. nauk/V.Ju. Shipicin. N. Novgorod, 2002. 26 s.
68. Sgibnev, D.V. Novyj sposob profilaktiki vnebol'nichnyh pnevmonij v voinskih kollektivah: avtoref. dis. ... kand. med. nauk/D.V. Sgibnev. N. Novgorod, 2004. 24 s.
69. Bahmudov, G.G. Gigienicheskoe obosnovanie optimizacii raciona pitanija pogranichnikov pri adaptacii k sluzhbe v kontrastnyh klimaticheskih uslovijah (Prikaspijskaja nizmennost' – srednegor'e): avtoref. dis. ... kand. med. nauk/G.G. Bahmudov. N. Novgorod, 2005. 25 s.
70. Averkin, D.A. Sovershenstvovanie sistemy profilakticheskih meroprijatij pri kishechnyh infekcijah v organizovannyh voinskih kollektivah, dislocirovannyh v zharkom klimate: avtoref. dis. ... kand. med. nauk/D.A. Averkin. N. Novgorod, 2005. 23 s.
71. Mihanoshina, N.V. Zabolevaemost' lic organizovannogo kollektiva v razlichnyh pogodno-klimaticheskih uslovijah Dal'nego Vostoka/N.V. Mihanoshina//Sanitarnyj vrach. 2013. № 9. S. 61–63.
72. Bobrov, N.I. Nauchnye osnovy zakalivanija organizma/N.I. Bobrov, O.P. Lomov, V.P. Tihomirov//Voenn.-med. zhurn. 1979. № 7. S. 25–29.
73. Kislicyn, A.N. Sravnitel'noe issledovanie adaptivnyh reakcij organizma cheloveka k uslovijam gornogo i morskogo klimata/A.N. Kislicyn. M., 2000. 19 s.
74. Rubin, A.D. Nauchno-organizacionnye osnovy optimizacii profilakticheskoj raboty s det'mi v uslovijah evropejskogo Severa Rossii: avtoref. dis. ... dokt. med. nauk/A.D. Rubin. SPb., 2009. 29 s.
75. Baklanov, P.Ja. [i dr.]. Geograficheskoe polozhenie Primorskogo kraja//Geografija Primorskogo kraja; pod red. G.A. Kakorinoj. 2-e izd. Vladivostok: Dal'press, 2000. 180 s.
76. Veremchuk, L.V. Vlijanie prirodno-jekologicheskih uslovij na kachestvo sredy obitanija cheloveka v Primorskom krae/L.V. Veremchuk, B.I. Chelnokovova//Zdorov'e naselenija i sreda obitanija, 2013. № 2. S. 4.
77. Bogomolov, A.B. Osobennosti techenija vnebol'nichnoj pnevmonii u voennosluzhashhih v sovremennyh uslovijah v zavisimosti ot struktury kontingentov i klimatogegraficheskih aspektov: avtoref. dis. ... kand. med. nauk/A.B. Bogomolov. SPb., 2010. 21 s.
78. Mihanoshina, N.V. Ocenka jepidemicheskogo processa vnebol'nichnyh pnevmonij v uslovijah Kamchatki/N.V. Mihanoshina, R.S. Rahmanov, N.N. Potehina//Zdorov'e naselenija i sreda obitanija. 2014. № 2 (2151). S. 31–34.
79. Nagaev, V.M. Adaptacija pogranichnikov k uslovijam Zapoljar'ja i sposoby ee optimizacii v nachal'nyj period sluzhebno-boevoj dejatel'nosti: avtoref. dis. ... kand. med. nauk/ V.M. Nagaev. N. Novgorod, 2000. 28 s.
80. Polevaja, A.L. Nauchnoe obosnovanie sistemnogo ispol'zovanija fizicheskih prirodnyh lechebnyh faktorov kurorta Anapa v reabilitacii detej, stradajushhih slizisto-gnojnym hronicheskim bronhitom: avtoref. dis. ... kand. med. nauk/A.L. Polevaja. Sochi, 2010. 24 s.
81. Homenko, S.N. Spravochnik po fizicheskoj geografii Sahalinskoj oblasti/S.N. Homenko. Juzhno-Sahalinsk: Sahalinskoe knizhnoe izdatel'stvo, 2003. 110 s.
82. Hizhnjak, Ju.Ju. Osobennosti techenija bronhial'noj astmy v uslovijah mussonnogo klimata Sahalina : avtoref. dis. ... kand. med. nauk/ Ju.Ju. Hizhnjak. Blagoveshhensk, 2009. 24 s.
83. Orlenok, V.V. Regional'naja geografija Rossii. Kaliningradskaja oblast'/ V.V. Orlenok, G.M. Fedorov. Kaliningrad: RGU im. I. Kanta, 2005. 258 s.
84. Barinova, G.M. Kaliningradskaja oblast'. Klimat/G.M. Barinova. Kaliningrawd: Jantarnyj skaz, 2002. 196 s.

85. Gigiena: uchebnik dlja med. vuzov; pod red. G.I. Rumjanceva. M.: GJeOTAR Media, 2015. 608 s.
86. Tarasov, A.V. Gigienicheskaja optimizacija zdorov'ja kursantov vysshego voennogo uchebnogo zavedenija pri adaptacii k morskomu klimatu (region Baltijskogo morja): avtoref. dis. ... kand. med. nauk/A.V. Tarasov. N. Novgorod, 2014. 23 s.
87. Fillipova, O.N. Gigienicheskie osnovy optimizacii pitanija natural'nymi kriogennymi produktami podrostkov s vysokoj fizicheskoj aktivnost'ju: avtoref. diss. ... kand. med. nauk/ O.N. Filippova. N. Novgorod, 2015. 23 s.
88. Dadali, V.A. Minornye komponenty pishhevyh rastenij kak reguljatory detoksikacionnyh i metabolicheskih sistem organizma/V.A. Dadali//Vestnik Sankt-Peterburgskoj gos. medakademii im. I.I. Mechnikova. 2001. № 1. S. 24–30.
89. Zaxar`in, G. A. Klinicheskie lekcii: prakticheskoe posobie / G. A. Zaxar`in. – Izd. 5-e. – Moskva : b.i., 1895. – Vy`pusk 1. Vvedenie v klinicheskie zanyatiya (3 lekcii). – 137 s.
90. Assman, D. Chuvstvitel`nost` cheloveka k pogode. L.: Gidrometeoizdat, 1966. 245 s.
91. Mezerniczkij, P.G. Medicinskaya meteorologiya / prof. P. G. Mezerniczkij; Gos. in-t med. klimatologii i klimatoterapii. - 2-e izd., ispr. i dop. - Yalta : 1937. - 348 s.
92. Remizov, N. A. Uchebnik medicinskoj meteorologii i klimatologii dlya vrachej i studentov / N. A. Remizov. - Moskva ; Leningrad. - 1934.
93. Chubinskij, S.M. Bioklimatologiya. M.: Medicina, 1965. 198 s.
94. Danishevskij, G.M. Akklimatizaciya cheloveka na Severe: (S ocherkom kraevoj patologii i gigieny`) / D-r med. nauk prof. G. M. Danishevskij. - Moskva: Medgiz, 1955
95. Veremchuk, L.V. Vliyanie prirodno-e`kologicheskix uslovij na kachestvo sredy` obitaniya cheloveka v Primorskom krae / L.V. Veremchuk, B.I. Chelnokovova // Zdorov`e naseleniya i sreda obitaniya, 2013. - № 2. - S. 4.
96. Baxmudov, G.G. Gigienicheskoe obosnovanie optimizacii raciona pitaniya pri adaptacii v kontrastny`x klimaticheskix usloviyax (Prikaspijskaya nizmennost` - srednegor`e): avtoref. dis. ... kand. med. nauk / G.G. Baxmudov. - N. Novgorod, 2005. - 25 s.

Index

A. V. Tarasov, R. S. Rakhmanov, *Marine Climate of Russian Coastal Territories*,
https://doi.org/10.1007/978-3-031-30951-9